Editors

**Andrej Démuth
Slávka Démuthová**

SPECTRUM SLOVAKIA Series
Volume 44

A Conceptual and Semantic Analysis of the Qualitative Domains of Aesthetic and Moral Emotions

An Introduction

Bibliographic Information published by the Deutsche Nationalbibliothek
The Deutsche Nationalbibliothek lists this publication in the Deutsche Nationalbibliografie; detailed bibliographic data is available in the internet at http://dnb.d-nb.de.

This publication has been peer reviewed.

Editors:	Andrej DÉMUTH
Slávka DÉMUTHOVÁ

ISSN 2195-1845
ISBN 978-3-631-90301-8	ISBN 978-80-224-2028-0
ePDF 978-3-631-90710-8
ePub 978-3-631-90711-5
DOI 10.3726/b21103

© 2023 Peter Lang Group AG, Lausanne
Published by Peter Lang GmbH, Berlin, Deutschland

© VEDA, Publishing House of the Slovak Academy of Sciences
Bratislava 2023

All rights reserved.
Peter Lang – Lausanne · Berlin · Bruxelles · Chennai · New York · Oxford

All parts of this publication are protected by copyright. Any utilisation outside the strict limits of the copyright law, without the permission of the publisher, is forbidden and liable to prosecution. This applies in particular to reproductions, translations, microfilming, and storage and processing in electronic retrieval systems.

www.peterlang.com	www.veda.sav.sk

This work was supported by the Slovak Research and Development Agency under the Contract no. APVV-19-0166 ("An Analysis of the Conceptual and Qualitative Domains of Aesthetic and Moral Emotions", and project VEGA No. 1/0120/22 ("Cognitive Aspects of Social and Moral Emotions").

Contents

A Conceptual and Semantic Analysis of the Qualitative Domains
of Aesthetic and Moral Emotions: An Introduction
Andrej Démuth, Slávka Démuthová 7

On Some Etymological, Grammatical and Contextual Reasons
for the Vagueness of the Concept of Beauty
Andrej Démuth, Slávka Démuthová, Yasin Keceli 39

The Possibilities of Studying Connotations of the Term "Beauty"
in a Natural Language
Slávka Démuthová 57

Considering the Emotion of Disgust in the Context of Terminology
and Contemporary Literature
Renáta Kišoňová 81

Spiritual and Theological Discernment of Good and Evil
Ľubomír Batka 101

The Relevance of Legal Intuitionism and Selected Moral Emotions
in Legal Thinking and Decision-Making Processes
Olexij M. Meteňkanyč 119

Notes on Contributors 165

A Conceptual and Semantic Analysis of the Qualitative Domains of Aesthetic and Moral Emotions: An Introduction

Andrej Démuth, Slávka Démuthová

Abstract. The Introduction aims to define moral and aesthetic emotions, and demonstrate the potential of the analysis of terms that are characteristic of moral and aesthetic emotions. We will attempt to clarify the various approaches (etymological studies, study of the use of terms in the ordinary language, clustering and the statistical analysis of the connotations of terms, semantic analysis of synonyms and opposites, conceptual analyses,...) that we deem to be beneficial for a systematic and detailed studies of basic terms to define moral emotions and aesthetic emotions, their schematic domains and possibly also the relationships between the studied terms and concepts (their position in the semantic fields).

Introduction

Emotions are an integral part of our lives. They are often considerably more important than we care to admit. With a slight amount of exaggeration, we could say, we are what and how we feel, rather than how we see ourselves or how others see us. After all, even our self-image provokes an emotional reaction from us, as much as our perception of how others see us. For all our lives, we are confronted by our emotionality and, to a certain extent, our emotionality is who we are (Kvajo, 2016).

And yet, emotions are often not the target of our intellectual interest. Quite the opposite. Not only do we frequently fail to correctly and clearly understand what emotions are and the function they fulfil in our lives, what they reveal and manifest, but we also often do not know how to cope with them – despite the fact that feelings and emotions are, just like our moods, omnipresent. We are always in some kind of emotional state – we could describe our emotionality as a sort of mental weather. And just like there is always weather, we are always in some sort of mood or emotional state, even if we cannot quite describe or name it because it appears somehow ambiguous. We are always in some kind of mood, even when we say we are not in the mood for anything (specific) (Heidegger, 1927). This very idea that our emotionality, in its most diverse forms, is such an integral part of our existence should be so vital and tempting that we should pay more attention to it. But most of the time, this is certainly not the case. We only consider emotions and mood to be mental colours that do not colour our worldview. We do not reflect on why we have this perception, what it means or whether it is possible to see the world even without colouring – which on occasion is quite substantial and in other circumstances is somewhat negligible.

Philosophers are often interested in emotions, if they are cognitive obstacles that distort the clarity and purity of rational thought and judgements. For aesthetes, they are obstacles to purely aesthetic judgements, which ought not to be obscured by emotions. For psychologists, emotions are substantial when they bother or threaten us, when we cannot cope with them or when they pose problems to others. Psychiatrists are interested in emotions in borderline cases, when they manifest as a pathological state, mainly they become the centre of their attention when they consider their management and suppression. Thus, we have psycho-pharmaceuticals, which tone down states of ecstasy, repress depression, regulate the cycles of bipolar disorders, and provide help to cope with other various undesirable emotional states, emotions and moods. There are also many courses available to teach us how to manage our emotions (to repress and manage anger, fear, stage fright and many others), how to suppress them or how to use them. "Over the last decades there has been an increased interest on the role of emotions in the multifaceted experience of work (Kluemper et al., 2013). Affective factors (e.g., emotions, feelings, mood etc.) have been recognized not only as pervasive in regulating and guiding human behavior (Von Scheve, 2012), but also directly related to individual and organizational well-being, performance and job satisfaction (Seo et al., 2010)" (Carminati, 2021). But do we even know why emotions exist and what use they serve?

The development of cognitive sciences and imaging technologies has given us a partial understanding of the physiological, neurobiological and neural correlates of specific emotions, but despite undeniable progress in our understanding, "the relative lack of objective methodologies [for the scientific study of] emotional phenomena limits our current understanding and thereby calls for the development of novel methodologies" (Gu et al., 2019). There is also a lack of integrated interdisciplinary research into emotionality which would allow us to understand not only how emotions are created, what affects them and what purpose they serve but also what they reveal to us in real life and what is their essence.

Homo as Rational Animal

One of the reasons that we tend to overlook or ignore our emotionality is the influence of (not only) Western philosophical tradition; we often believe that reason and rational thinking are what is important to us. This is not surprising, considering that man has been defined as a rational animal since the era of Aristotle, and so reason ought to be the essential thing that characterises us and makes us different from other living beings. We call our species Homo sapiens (wise man) to highlight our rationality and wisdom, and this is what makes us different from other animals or even our ancestors (such as Homo neandertalis). Thus, in many concepts, reason is the essence of our being. Whether we call it our spirit, memory (Aristotle's passive intellect) or intellect – reason is what we unknowingly identify with our own identity.

We consider reason (as opposed to emotions) to be more-or-less stable, virtually unchanged over the course of time. The laws that govern our thoughts, logic, prisms and the like stay the same from when we are born until our final thoughts. We believe that our beliefs – the axioms of our thoughts – do not change to any great extent and remain constant; they are a sort of Archimedean point of thinking. And even if they do change, it is more of a change of the style of the ship of Theseus rather than a sudden and complete rupture, paradigm shifts and changes of our views. Thus, we also see ourselves as more-or-less consistent and stable in our thoughts, values and partially in the opinions we hold. The constant nature of reason not only secures our stability and the identity of an individual within their self-reflection. The universality and timelessness of reason allow people to share knowledge with others through various environments or even entire periods. The use of the same rationality allows us to communicate with others and be understood, and the stable and unchanging nature of forms of thought proves the timelessness of the ideas we come across when reading the works of old writers. We even assume that rational beings, provided that they think purely rationally and employ the same information, will come up with more-or-less the same results, regardless of their historical and geographical points of departure (Démuth, 2013). After all, the assumption of universality and the historical nature of rationality are part of the foundations of science; they allow its findings to remain valid and our thoughts to maintain a degree of consistency and super-individuality.

Emotions as a Synonym of Instability and Subjectivity

On the other hand, emotions are often perceived as sudden and fleeting. They occur to various degrees and with various intensities, and their duration is highly inconsistent. Some burn out as suddenly as they appear, as if they have exhausted all the energy or oxygen in the space they have occupied. Some are more permanent and may last minutes or hours, and then there are others that are relatively stable (such as deep grief after the loss of a loved one, or depression), which is what makes them so burdensome – they may last months or even years. If, in the Platonian spirit, reason could be compared to the Sun, which is permanent and gives us light, emotionality may be compared to the conditions that cause changes in the weather. The Sun even shines during a cloudy autumn day, but due to the fog, cloud and high humidity in the atmosphere, its rays do not reach us and we feel that everything is cold and miserable. This is a typical feature of emotions.

Although they are temporally unstable and can vanish after a very short time, they substantially change our perception of things. While reason tries to gain an objective understanding, free of any subjective considerations, emotions are principally subjective. Emotions subjectively colour our view of the world; they distort it and highlight our own position (Démuth, 2003). Furthermore, what becomes available through one emotion may be concealed or distorted by another. Thus, emotional states are often perceived as a synonym for irrationality – for something that is neither rational nor objective. They are an expression of a situation-dependent subjective perception and experience of reality, not of something stable and unchanging. What a person might perceive through the prism of their emotions, another person might perceive quite differently. What is more, as our emotions change, the same reality could be perceived quite differently than it was in a previous emotional state.

The problem of emotionality does not merely lie in volatility (instability) and subjectivity. An equally serious problem is the depth and complexity of emotions and our level of self-awareness. Many feelings and emotions appear at the most fundamental levels of consciousness, and it is presumed that they not only exist in small children but also in embryos and other animal species. This leads us to the following question: what are emotions and where and how can we meet them in terms of ontogeny and phylogeny.

What Are Emotions? An Attempted Definition?

Classical psychological dictionaries (such as Reber, 1995) point out that there are only few terms that are as ambiguous and hard to define as emotionality. "Most textbook authors wisely employ it as the title of a chapter and let material presented substitute for a concise definition. The term itself derives from the Latin *emovere* which translates as to move, to excite, to stir up or to agitate. Contemporary usage is of two general kinds: (1) An umbrella term for any number of subjectively experienced, affect-laden states, the ontological status of each being established by a label whose meaning is arrived at by simple consensus. This is the primary use of the term in both the technical and the common language. It is what we mean when we say that love, fear, hate, terror, etc. are emotions. (2) A label for a field of scientific investigation that explores the various environmental physiological and cognitive factors that underlie [these] subjective experiences" (Reber, 1995, 246).

The Theories of Emotions

Furthermore, the history of philosophy and psychology has presented many diverse and sometimes even contradictory theories and views on what are emotions. For instance, Darwin (Darwin, 2007 [1872]) assumed that emotions exist as adaptive patterns of behaviour that allow organisms to successfully survive. He assumed that there are certain established patterns of behaviour (in us and other species) driven by three basic principles: "the principle of serviceable habits", "antithesis" and "expressive habits" (Darwin, 2007 [1872]), whose teleological explanation was not quite clear. These forms of behaviour have outward expressions, which he believed to be expressions of emotions. According to him, emotions allow organisms to react to the stimuli around them, which consequently helps them to improve their chances of success (to cooperate) and thus their survival. The problem of such a broadly based evolutionary theory (in spite of its numerous modifications – Gaulin, McBurney, 2003) is that almost everything may be understood as an established form of behaviour (feelings, reflexes up to complex high-level feelings) and that it fails to clarify why these forms developed in this specific way and not another and how they are different from feelings, instincts and reflexes. At the same time, it fails to clarify why intentional emotional expressions influence the way they are experienced (Facial-Feedback Theory of Emotion – Tomkins, 1962 in Adelman, Zajonc, 1989).

James-Lange's theory of emotion (James, Lange, 1922) assumes that emotions are physiological reactions to a perceived stimulus. Subsequently, we subjectively experience these physiological processes in the form of, what we call, emotions (Dewey, 1894, Barrett, 2017). Therefore, if we perceive an impulse that we interpret as danger, the organism reacts to it with a cascade of physiological processes (a change of heart rate, breathing, galvanic skin response, etc.), which we subjectively evaluate as, for example, "fear". Hence, they are subjectively experienced "biological judgements of a situation" that trigger these feelings and physiological changes.

A sophisticated version of this somatic theory of emotion is Damasio's theory of somatic markers. According to this theory, emotions, as defined by Antonio Damasio, are changes in the state of body and brain in response to impulses (Damasio, 2008 [1994]). Physiological changes (such as muscle tone, heart rate, endocrine activity, facial expression, etc.) are bodily states and are transferred to emotions by the CNS. The specific emotion will tell the individual something about the stimulus they have just encountered. Their emotions and corresponding physical changes, which he calls "somatic markers", are associated with specific situations and their past consequences. If somatic markers and the emotions they provoke are consciously or subconsciously associated with past consequences (such as feelings of happiness or reward), they influence decision-making to the advantage of one behaviour rather than another, that is associated with a negative result. In fact, Damasio distinguishes processes without any conscious mechanism of deduction (the need to eat evolved as a result of a drop in glucose levels and the consequent feeling of hunger), which provoke an automatic reaction (reflexes), decision-making with a conscious deduction mechanism and personal bias (emotions in their broader sense), and decision-making with a conscious deduction mechanism without personal bias, which is characteristic of thought and theoretical reasoning.

Cannon-Bard theory rejects James-Lange's somatic understanding of emotions. Cannon (Cannon, 1927) realised that a number of physiological changes occur without any emotional expression or even after something is experienced (e.g., one is spooked and only then there is a change in heart rate). Thus, he assumes that emotions are mental reactions to impulses that cannot be reduced to mere physical expressions. They may co-occur, but one does not trigger the other.

Schachter-Singer's two-factor theory of emotion also takes both physiological excitement and emotional experience into consideration (Schachter, Singer, 1962). The theory suggests that emotions are made up of two factors: a physiological factor and a cognitive factor. Physiological excitement is interpreted in a cognitive context to create emotional experience. According to this theory, the key constituent is the physiological reaction of an organism and the cognitive understanding of the situation (the physiological processes experienced). And only this cognitive assessment allows us to experience emotions. This would explain why different people experience different emotions in a situation that from the outside seems to be the same. Their assessment of the situation is not the same.

This resembles Lazarus' theory of cognitive appraisal (Lazarus, Folkman, 1984). This theory proposes that an idea (the cognitive appraisal of a situation) must take place before the physiological (or emotional) experience of emotions. Discovering a stimulus leads to its appraisal, and emotions only appear as a reaction to this appraisal.

There are also theories that emotions are actually judgements (Solomon, 1993), action tendencies (Frijda, 1988), a part of broader heuristics (Chaiken, Liberman & Eagly, 1989; Taleb, 2001), a part of a model of affect infusion (Forgas, 1995) or two-level perceptions of our own state (Prinz, 2004). Clearly, according to these theories, emotions are a consequence of the perception and interpretation of our own state, situation and desire.

A situational perspective of emotion (Griffiths, Scarantino, 2005) sees emotions as a product of an organism that monitors its environment and observes (intentional and unintentional) the reactions of other organisms. Similarly, the affective events theory (Weiss, Beal, 2005) assumes that emotions are based on communication and aim to generate a degree of cooperation between individuals. In addition, there are also modern neurophysiological (Cacioppo, 1998) and neurobiological theories of emotionality (LeDoux, 1996; Adelmann, Zajonc, 1989; McIntosh, Zajonc et al., 1997), as well as genetic or even lucid philosophical explorations (de Soussa, 1987; Goldie, 2000, 2002; Ratcliffe, 2005)...

Philosophy and Emotion

In philosophy, we mostly distinguish emotions, perceptions and feelings, emotions and assessments, and emotionality and motivation, but even in this field the distinctions are not quite unified. We keep asking if emotions are always intentional: (are emotions object-directed, and if they are, can they be appropriate or inappropriate for their objects?), whether they always and inevitably have a subjective phenomenal content (do emotions always involve subjective experiences, and if so of what kind?). We study them as evaluative perceptions, as evaluative feelings or as patterns of salience. We reflect upon whether they are always accompanied by a physical response or rather if their essence lies in experiences (the problem of enactivism). Despite the rationalist tradition, we ask ourselves whether they are not, in fact, rational or at the very least if they can be subject to rational analysis (de Sousa, 1987; Scarantino, de Sousa, 2021).

"The simplest theory of emotions, and perhaps the theory most representative of common sense, is that emotions are simply a class of feelings, differentiated by their experienced quality from other sensory experiences like tasting chocolate or proprioceptions like sensing a pain in one's lower back" (Scarantino, de Sousa, 2021). Ever since the time of the Ancient Greeks, we have believed that there are certain primitive contents that do not include any other components. These contents correspond to sensations, which we understand as what is detected by receptors (a physical process). However, the whole thing is complicated by the fact that we cannot directly perceive what takes place in the receptors. We become aware of these contents, and we are only able to interpret them in our consciousness (a mental process). Thus, the subject of this conscious interpretation is a perception. A perception may be simple or complex, but it mostly provides meaning – it interprets the reality that we come across (the perception of red). On the other hand, feelings are also perceptions (reflections) of one's own experience. They also refer to a conscious process, a sensory content (e.g., the feeling of warmth), an affective state (e.g., feeling good), to one of the three dimensions of emotion in Wundt's theory, or to an ambiguous belief – "a hunch", as we commonly but inaccurately say (Reber, 1995, 284). On the other hand, we understand emotions as a resultant, relatively complex set of data that the subject perceives as self-states or as states of their relationship to others or to their self. Unlike feelings, emotions may be subconscious, and we may not be aware of them.

Just as we have countless different types of feelings, there is a plethora of emotions. Philosophers dispute whether emotions are natural, discrete entities, whether they are the product of a social constructivism, or whether they are not discrete entities but simply contents on a continuum. This continuum may be steady in principle, but it may also contain supporting points – dimensions that are a sort of ground zero.

"Basic emotion theory has been very influential for more than half a century, providing inspiration for interventions in psychopathology (Saarimaki et al., 2016; Celeghin et al., 2017; Williams, 2017; Hutto et al., 2018; Song and Hakoda, 2018; Vetter et al., 2018; Wang et al., 2018). Theories about basic emotions originated from ancient Greece and China (Russell, 2003). Current basic emotion theory started with Darwin (Darwin, 2007 [1872]) and Ekman (Ekman, 2003), and later (Tomkins, 1962), subsequently followed by Ekman (Ekman, 1984), and Izard (Izard, 1977), then by many current psychologists (Ortony and Turner, 1990; Panksepp, 2007; Scarantino and Griffiths, 2011; Gu et al., 2016, 2018; Saarimaki et al., 2016; Hutto et al., 2018). Basic emotion theory proposes that human beings have a limited number of emotions (e.g., fear, anger, joy, sadness) that are biologically and psychologically 'basic' (Wilson-Mendenhall et al., 2013), each manifested in an organized recurring pattern of associated behavioral components (Ekman, 1992; Russell, 2006)" (Gu et al., 2019). Basic emotions (if there are such things) are most probably universal, and they perform different functions and possibly come from different neural structures or neurotransmitters; they serve as beacons in our mental world, allowing us to map the other corners of our emotional/mental world. Naturally, we need to have separate terms that refer to these basic emotions and that differentiate them from other contents. And herein lies the essence of the problem.

Just as we are able to perceive a limitless number of colours of all varieties but lack a sufficient number of individual terms to differentiate between the perceived shades, it appears that there are also a limitless number of emotional affective contents, even though we do not have a sufficient large number of terms that would allow us to differentiate between them with sufficient accuracy. This means that our emotional and affective world is significantly more fractured than we are able to describe through language.

Despite the many discussions on the role and purpose of philosophy, it appears there is a general consensus (since Wittgenstein, at least in the analytical philosophical tradition) that one of the key roles of philosophy should be to clearly and accurately map out the real and fictitious world using conceptual schemes and terms. This means that one possible contribution of contemporary philosophy to the question of emotions may be to find mechanisms that answer the following question: How are emotions different from each other, and from things that are not emotions?

Objective

The primary aim of the conference proceedings is to define moral and aesthetic emotions. Through a conceptual and semantic analysis of the basic terms, we wish to define (normative function) and clarify (descriptive function) the meaning and scope of the studied terms (Laurence, Margolis, 2003), including *aesthetic experience, aesthetic emotions, moral experience, moral emotions, ...*

The second aim of the presented studies is to demonstrate the potential of the analysis of terms that are characteristic of moral and aesthetic emotions. We will attempt to clarify the various approaches that we deem to be beneficial for a systematic and detailed studies of basic terms to define moral emotions and aesthetic emotions, their schematic domains, and possibly also the relationships between the studied terms and concepts (their position in the semantic fields).

Attempts to Clarify Terminology

Philosophers have dealt with the topic of moral and aesthetic emotions since ancient times. Plato, Aristotle as well as the stoics (Seneca, etc.) scrutinised the descriptions of various feelings that accompany the experience of beauty, injustice, anger and other emotions. However, mostly they did not truly attempt to define these emotions.

What Are Moral Emotions?

One of the first contemporary systemic attempts to define the essence of moral emotions as a separate field of study was the work of Jonathan Haidt (Haidt, 2003) in *The Moral Emotions*. Haidt and many others who followed him (Pizarro, 2007, Tagney, Stuewig, Mashnek, 2007) understand moral emotions to be emotions which occur with the perception of a real or potential action, situation or phenomenon and an evaluation of whether it is moral or immoral. Although this definition is (tauto)logically convincing, it is not very helpful. It assumes that the reader has a sufficient understanding of what constitutes moral or immoral actions, what emotions are and what separates them from other states or feelings. But this is not always the case. What may appear to be moral from one perspective may appear to be immoral or morally ambivalent from another.

For this reason, Haidt attempts to define the morality of emotions more accurately and categorically with the proposal that they "are linked to the interests or welfare either of society as a whole or at least of someone other than the judge or agent" (Haidt, 2003). Although this definition helps us to understand what the author means by the attribute "moral", it seems that this definition is not entirely sufficient, nor complete. The concept of the welfare of the whole is still too unclear and it is questionable whether all moral emotions must be related to an interest in the welfare of a society or its individual members. It is questionable whether it is necessary for another person to be present for me to feel ashamed of myself (for someone else?), or if it is rather a feeling solely connected to myself (Scheler, 1957). Whether grief or regret must always be felt for someone else or for society as a whole. However, it cannot be disputed that for several moral emotions, primarily we do not care about the welfare of society as a whole (such as feeling hatred towards something) but rather about problems that only affect a part of society.

Haidt realises that emotionality is a highly fractured and complex field, which is why he does not try to describe it using definitions that would exactly define all moral emotions. After all, philosophers, psychologists and moralists are still discussing the number of moral emotions, what they are, whether there are any fundamental emotions, from which the rest are merely derived or made up of, or whether there are some basic characteristics that are typical of some groups of emotions. Haidt believes that "the principal moral emotions can be divided into two large and two small joint families. The large families are the 'other-condemning' family, in which the three brothers are contempt, anger, and disgust (and their many children, such as indignation and loathing), and the 'self-conscious' family (shame, embarrassment, and guilt)... [T]he two smaller families [are] the 'other-suffering' family (compassion) and the 'other-praising' family (gratitude and elevation)" (Haidt, 2003).

Gray and Wegner adopted a different approach to defining moral emotions. They both believe that "people divide the moral world along the two dimensions of valence (help/harm) and moral type (agent/patient). The intersection of these two dimensions gives four moral exemplars – heroes, villains, victims and beneficiaries – each of which elicits unique emotions. For example, victims (harm/patient) elicit sympathy and sadness. Dividing moral emotions into these four quadrants provides predictions about which emotions reinforce, oppose and complement each other" (Gray, Wegner, 2011).

There are also other strategies for the definition of morality (e.g., by what is good or bad – Prinz 2009) or even for differentiating between moral and immoral emotions.

Many of these definitions strive to be precise and complex (e.g., "Moral emotions are a variety of social emotion that are involved in forming and communicating moral judgments and decisions, and in motivating behavioral responses to one's own and others' moral behavior." (Tangney et al., 2007), or see *An Attributional Analysis of Moral Emotions* (Rudolph & Tscharaktschiev, 2014)); yet they often appear to be incomplete or inaccurate. Hence, some philosophers suggest we abandon the effort to define what makes moral emotions moral, and instead focus on the context in which they occur.

What Are Aesthetic Emotions?

The situation related to aesthetic emotion is equally complicated. Winfried Menninghaus et al. (2019) tried to define aesthetic emotion. They concluded that they are emotions that involve "an aesthetic evaluation/appreciation of [an object]" and that they also appreciate qualities other than beauty; they are associated with subjective pleasure or displeasure and so forth. According to Menninghaus, aesthetic emotions are categorically different from other emotional states, even though terms for aesthetic feelings are created through an overlap of aesthetic evaluation with the accompanying feelings or, on the contrary, by the overlap of emotional meaning with prototypical terms for aesthetic appreciation. The fact that they overlap with affective states suggests the presence of moments of novelty or, on the contrary, the presence of familiarity, excitement and pleasure, as well as several cognitive aspects; the presence of behavioural emotional expressions, neurophysiological activity of the reward system and several other moments (such as reinforcing the intensity with negative emotions and states). Despite efforts to define aesthetic emotions as complexly and as precisely as possible, here it also seems that defining this type of emotion in other than a tautological way (i.e., that they are emotions that occur when perceiving and aesthetically evaluating an object) is not entirely sufficient, since it sparks a broad (and at times highly critical) discussion and rejection of the argument (e.g., Skov, Nadal, 2020), or it results in further attempts to find a new definition for aesthetic emotion (Fingerhut, Prinz, 2020; Menninghaus et al., 2020; Conte, Hahnel, Brosch, 2022). As demonstrated by Leonid Perlovski (Perlovski, 2014), it is not only the definition of emotions but also the definition of what "aesthetic" means that is problematic.

It is just as problematic, or even more so, to define what we mean when we speak about the concept of emotions. Are they discrete psychosomatic states that we are fully aware of, that are associated with a particular type of emotional mental experience of an individual, or are they affective states of an organism? Are they states of discrete, distinguishable feelings, or are they rather continuums of affective states separated by the various degrees of saturation within the various dimensions of these feelings and states (Démuth, 2017)? In order for them to exist, does the subject need to be aware of them?

A Preliminary Definition of the Moral and Aesthetic Emotions

Although scientists have been waging a 100-year-long war on the nature of emotions and their main characteristics and even though we do not have any generally accepted definition of what is emotionality and what makes moral emotions moral and aesthetic emotions aesthetic, our study is based on the assumption that emotions "are internal states of [an] organism that are modulated by neuromodulators, and these internal states are externally expressed as certain stereotypical behaviours, such as instinct, which is proposed as ancient mechanisms of survival" (Gu et al., 2019). Emotions can therefore be understood as subjectively experienced, affect-laden states that mostly give the subject information on the state of their own organism, but they also represent reactions to internal and external impulses, depending on the evolutionary and individually acquired knowledge.

Moral emotions should especially be understood as affective states that provoke, accompany or result from ethical or moral (conscious or subconscious) evaluation, or as states that occur when we perceive and evaluate a situation or a phenomenon within the dimensions of "desirable – undesirable" or "good – bad", not only from the perspective of the benefit to the individual but also considering the individual's coexistence with others. Analogously, by aesthetic emotion we mean the wide array of affective states that accompany aesthetic perception and evaluation of an object, phenomenon or state in the dimensions: "liking – disliking", a feeling of interest, pleasure or other aesthetic evaluation of an object, regardless of what the object is, what purpose it serves or who created it (Kantian understanding of the criteria of pure aesthetic judgement). We believe that in spite of the apparent incompleteness of this definition, our understanding of both types of emotions will be, thanks to intuition, sufficiently clear, relying on the characterisation in various works

(Eisenberg, 2000; Haidt, 2003; Tangney et al., 2007, Frijda et al.,1989; Keltner and Haidt, 2003; Silvia, 2005, 2009; Scherer & Coutinho, 2013; Perlovsky, 2014; Menninghaus et al., 2019).

How to Study Moral and Aesthetic Emotions?

Moral and aesthetic emotions may be studied through various approaches and processes. One is the study of mental and behavioural correlates of the emotions or states. In the case of mental expressions, we are mostly interested in what the individual experiences when they feel an affective state and what separates it from a different emotional state. This approach is especially typical of psychological studies and phenomenology or the hermeneutics of emotional experience.

An example of such phenomenological-hermeneutic research is Heidegger's analysis of fear and anxiety in his *Sein und Zeit*. Heidegger mainly analyses what the various objects that, for example, trigger fear have in common, i.e., how we understand them and why we are interested in them. He then analyses the characteristics of certain emotional experiences, for instance, what characterises the experience of fear, what we feel and what is the purpose of this experience. The final question asked by phenomenological analysis of emotional experience is what is the point of the experience. It is not just that the point/purpose of a given type of experience can be studied from an evolutionary standpoint, i.e. from the perspective of the existence of the species (this is studied by evolutionary psychologists and evolutionary biologists) but also from the standpoint of the meaning of the content of such an experience for the individual. Heidegger (and, in fact, Soren Kierkegaard before him) established a remarkable tradition of the phenomenological-hermeneutic-existential study of emotions, which has been very popular in philosophical studies (Steinbock, 2014 – moral emotions of self-givenness (pride, shame, guilt), possibility (repentance, hope, despair) and otherness (trust, loving, humility); Scheler – shame), etc. The popularity of this approach that focuses on content and experience is also documented by the whole series:

The Moral Psychology of Emotion – Shame (Fussi, Rodogno, 2023, Seok 2017), Love (Pismeny, Brogaard, 2022), Hate (Birondo, 2022), Boredom (Elpidorou, 2022), Amusement (Robinson, 2021), Anger (Cherry, Flanagan, 2017), Disgust (Strohminger, Kumar, 2018), Sadness (Gotlib, 2017), Pride (Carter, Gordon, 2017) Envy (Protasi, 2022), Hope (Bloser, Stahl, 2019), Regret (Gotlib, 2022), Gratitude (Roberts, Telech, 2019), Curiosity (Inan, Watson, Whitcomb, Yigit, 2018). Contempt (Mason, 2018), Admiration (Archehr, Grahle, 2018), Guilt (Cokelet, Maley, 2019), Trust (Collins, Jovanovič, Alfano, 2023) published by Rowman & Littlefield International.

The analysis of behavioural expressions of affective states is often based on the Darwinian thesis on the particular outward expression of emotion. Facial expressions could become a secondary subject for studies into the comprehension of emotions, as carried out by Paul Ekman and Wallace (Ekman, Friesen, 1971).

Yet another approach is the analysis of neural correlates that occur in different affective states. This analysis reveals which brain centres and sections of the CNS are active during the different types of affects and feelings as well as which brain centres, structures and patterns contribute to how we feel and experience the individual emotions. Possibly the best-known and most controversial example of such a study is the neuro-aesthetic study by Kawabata and Zeki (2004), which sought to find the brain centres of beauty, i.e., the areas that are active when perceiving and evaluating beautiful objects. It is possible to study the neural correlates of other emotions through this approach, such as anger (Denson et al., 2009), guilt and shame (Bastin et al., 2016), injustice (Stallen et al., 2018) or admiration (Immordino-Yang et al., 2009), and many more.

That being said, this work does not focus on the neural correlates of emotions but instead on how we understand and characterise them using language. The typical trait of most emotions is that they provide information (on the state of the organism or on situations and phenomena around it) both to the individual and to the surrounding individuals on their state and intentions. Thus, emotions are an important tool for communication, and the expression and accuracy of the communication message are their integral parts. This can be examined by the above-mentioned behavioural analysis of, for example, facial expressions that relate to individual emotions and our ability to comprehend them in the cultural context or even universally, throughout various cultural or social contexts. Or, on the other hand, it is also possible to examine how we, as individuals, understand emotions, their similarities and differences, which is reflected in how we classify and categorise them in our natural language.

Words – A Window into the Mental Space of the User

The subject of our analyses will be a study of the terms that we commonly use to classify emotions and affective states, how we group and separate them and what mutual links, we assume, exist between them. However, this perspective also offers many varied approaches that either focus on a semantic or conceptual analysis of emotional concepts and terms. First we will analyse the etymological roots of the concepts and terms.

Etymological Studies

In their study that serves as a gateway to the present study, Andrej Démuth, Slávka Démuthová and Yasin Keceli examine the etymological roots of the term "beauty" in various European languages. They highlight that although beauty is one of the key terms used in aesthetics and art, that appears in every culture, it seems that of all the aesthetic concepts, it is one of the least cohesive and most ambiguous. One reason is that in different and even closely related languages, the term refers to different semantic dimensions and meanings. This is documented in the well-known work by Crispin Sartwell, *Six Names of Beauty* (Sartwell, 2004), where the author highlights the different accentuations of the various levels of meaning of the term "beauty" in English (Beauty as an object of longing), Hebrew (Yapha as glow, bloom), Sanskrit (Sundara – holiness), Greek (to kalon – idea), Japanese (Wabi-sabi – perfection in imperfection) or in Navajo (Hozho – health, harmony). Beauty (or the feeling of it) is not only an umbrella term, but it also describes several levels of aesthetic appraisal that are often significantly discrete. Thus this term often refers to various denotations in different languages. And more. In the minds of different users of the very same natural language, this term is often understood on various semantic levels – it may evoke slightly different aspects in each user of the term. This frequently leads to inaccurate references, to vagueness and uncertainty in the use of this term. In other words, users of an umbrella term often imagine quite different contents, which frequently leads to misunderstandings and a relatively minor degree of agreement that could claim general consent and a binding nature for aesthetic judgements.

The ambivalence and uncertainty of the meanings can be minimised through additional shaping of terminology, "carving out" the non-original meanings or shelving the idea that a term can be used as an umbrella term for various other concepts. The study of the etymological roots of the individual concepts as well as efforts to coin a dictionary definition of their meanings requires a belief in the existence of an objective meaning of a term (which is given by a unified denotatum) and its adoption into the language in a historical (original or contemporary) form. Ludwig Wittgenstein (1921), who adopted this attitude to language in his Tractatus period (*Tractatus Logico-Philosophicus*), later reached the conclusion that the meaning of a word is not given by its positive definition but rather by how it is used in the language ("The meaning of a word is its use in the language" (Wittgenstein, 1921, 43). Thus the meaning of a term is ultimately always dependant on the way the language is used as well as the acceptance of that specific use by other users.

Hence, we should focus on the various experiences (literal, metaphorical and subjective) that evoke the use of concepts and terms that we use to convey meanings.

The Study of the Use of Terms in the Natural Language

In their book *Ancient Anger. Perspective from Homer to Galen*, published within Yale Classical Studies, Susanna Morton Braund and Glen W. Most (Braund, Most, 2003) presented an approach that attempted to map out the use of concepts and images for expressing different forms of anger in ancient literature. In their collection of studies on the topic of various aspects and forms of anger, we find images of Iliadic anger, angry bees, wasps and jurors, the rage of woman, your mother nursed you with bile, *Thumos* as the masculine ideal and social pathology in ancient Greek magical spells, Boy Achilles' diet, An ABC of epic *ira*: anger, beasts, and cannibalism... Similarly, we may map out the use of any randomly selected term that refers to an emotion or affective state in the Bible, Greek, Roman or Christian literature or anywhere else.

Studying dictionary definitions and the use of words in the corpora of the given language may allow us to identify the basic coordinates and all the possible extended uses of a term in a language, but it will not enable us to gain an insight into the everyday use of the term by a common user. To do so, it appears it is best to map out the connotations of a term in different common uses. This is why we study the language used to express, for example, feelings of anger, as well as metaphors that reflect the internal states when an individual experiences anger (Alia-Klein et al., 2020).

Both the literal and metaphorical uses of expressions allow us to map the semantic world of terms by looking at what they are associated with, be it in the collective or individual experience of the user. In fact, each term is associated with a subjective meaning and context in the mind of a user. Finding connections between these contextual and connotative meanings will help us better orientate ourselves in the semantic space and fields in the mind of a user.

A Statistical Analysis of the Connotations of a Term

In her study entitled *The Possibilities of Studying Connotations of the Term "Beauty" in a Natural Language*, Slávka Démuthová presents various theoretical approaches to the study of the use of emotion-labelling terms through their connotations. She assumes that an analysis of their frequency, a multifactor correlation analysis of connotations and a semantic differential of concepts, will help us better understand the fundamental structure and dimensions of the semantic levels or fields that the concepts occupy. Her work discusses the problem from a theoretical standpoint rather than that of a practical application.

However, the approach has been practically implemented in the published works of Démuth, Démuthová, Keceli (2022a; 2022b), Beermann et al. (2021) or Schindler et al. (2017) as well as in the prepared study by Andrej Démuth, who will try to focus on this type of statistical-analytical and semantic-differential approach to the concept of anger.

Semantic Differential

Semantic differential is a method designed for discovering subtle – subjectively specific attitudes and relationships to connotative meanings in stimulus words. It helps to identify subjective specificities in the understanding of concepts, as well as the level of agreement (clarity) or disagreement (vagueness) with a concept in the studied population. If the variance of the semantic differential equals zero, all language users have an identical understanding of the term. If the semantic differential is high, we speak about a very high rate of internal vagueness. "After its introduction in the 1950s, the semantic differential (SD) received enormous attention among psychologists and other social and behavioural scientists. Especially during the 1960s, its methodological details and possible fields of application were broadly and critically discussed" (Ploder, Eder, 2015, 563). Therefore, interest in this methodology declined in the 1970s, but it is still a standard psychological and sociological method that enables us to study individual attitudes to stimulus words and to discover the internal cohesion between the individual meanings in the minds and attitudes of various language users. For this reason, it is often a part of psycholinguistic research.

Synonyms and Opposites

In her text, Renáta Kišoňová presents yet another approach to the clarification and definition of terms used to label emotions through an analysis of their contents. There are two methods used to obtain meanings: the direct a priori definition of concepts; and the indirect argument. Kišoňová attempts to define these terms by analysing their synonyms (taken from dictionaries and common users – the direct approach) and their opposites (the indirect approach). By synonyms, she means not only words with identical meaning but also words with related meanings, i.e., those which exhibit certain differences besides a shared aspect of meaning. Through a conceptual analysis of the synonyms of the term "admiration", she has outlined the borders and plasticity of the semantic space of the term and has suggested its various semantic levels. She has adopted a similar approach to the analysis of synonyms of the term "disgust". She has studied both the selected terms and their associated semantic spaces that are delimited by synonym families, as opposites. She assumes that the meaning of a term cannot only be directly revealed but also indirectly – by illustrating the meaning of its opposites.

Conceptual Analysis

"In general, we may say that a conceptual analysis is a complex conceptual method, whose aim is to reveal, describe, or even create the relationships between the concepts of a certain conceptual system. The point of a conceptual analysis is often to find a concept B to replace concept A. It has the same meaning as concept A, but at the same time, the method used to identify the meaning through concept B is better, more accurate or transparent for various reasons – for example, with regard to the observed cognitive aims, term B more accurately sets out the conditions that the object must fulfil to be included in the extent of concept B, and thus also in concept A" (Zouhar, 2016, 413). In his study *Spiritual and*

Theological Discernment of Good and Evil, Ľubomír Batka reflects on the use of the concepts of good and evil, or sin, blame, remorse, through the prism of the person who evaluates the situation. He highlights that the terms used are not only associated with their content but also with the evaluator of the phenomenon and what it is compared to.

Olexij Mychajlovyč Meteňkanyč uses the same approach. His study provides a philosophical analysis of feelings of justice and injustice and their causes, but more frequently their consequences. He analyses the question of moral intuitionism and the understanding of ethical attitudes as moral emotions (moral sense). He highlights that feelings of injustice are not only a mere affective indicator of the existence of the moral attitude of the evaluator but that it is frequently the driving force behind societal changes or changes in the legal and moral norms of a society. Our feelings lead to changes in our actions, and changes in our actions lead to a change in the conditions around us. It is in this sense that feelings triggered by or aimed at others may be considered social or even moral emotions that lead to a search for a better world.

Conclusion

The present studies do not intend to provide a detailed or complex analysis of the selected aesthetic or moral emotions. They only attempt to outline the methodological approaches that could lead to a more profound and complex understanding of these types of emotions through an analysis of their terms in natural languages. The presented texts were created at the Centre for Cognitive Studies at the Department of the Theory of Law and Philosophy of Law, Comenius University in Bratislava, and at the Department of Psychology at Faculty of Arts, University of Ss Cyril and Methodius in Trnava as a part of project APVV 19-0166 "An Analysis of the Conceptual and Qualitative Domains of Aesthetic and Moral Emotions", and project VEGA No. 1/0120/22 "Cognitive Aspects of Social and Moral Emotions", and are pilot studies for further systematic research topics within the project.

References

Adelmann, P., K., Zajonc, R., B. (1989). Facial efference and the experience of emotion. *Annual Review of Psychology,* 40, 249–280. https://doi.org/10.1146/annurev.ps.40.020189.001341.

Alia-Klein, N., Gan, G., Gilam, G., Bezek, J., Bruno, A., Denson, T. F., Hendler, T., Lowe, L., Mariotti, V., Muscatello, M. R., Palumbo, S., Pellegrini, S., Pietrini, P., Rizzo, A., & Verona, E. (2020). The feeling of anger: From brain networks to linguistic expressions. *Neuroscience and Biobehavioral Reviews,* 108, 480–497. https://doi.org/10.1016/j.neubiorev.2019.12.002.

Archehr, A., Grahle, A. (eds.) (2018) *The Moral Psychology of Admiration.* Rowman & Littlefield Publishers.

Barrett, Feldman, L., (2017). *How Emotions Are Made: The Secret Life of the Brain.* New York: Houghton Mifflin Harcourt.

Bastin, C., Harrison, B., J., Davey, C., G., Moll, J., Whittle, S., (2016, Dec.). Feelings of shame, embarrassment and guilt and their neural correlates: A systematic review. *Neuroscience & Biobehavioral Reviews,* 71, 455–471. https://doi.org/10.1016/j.neubiorev.2016.09.019.

Beermann, U., Hosoya, G., Schindler, I., Scherer, K., R., Eid, M., Wagner, V., Menninghaus, W. (2021) Dimensions and clusters of aesthetic emotions: A semantic profile analysis. *Frontiers in Psychology,* 12, 667173. https://doi.org/10.3389/fpsyg.2021.667173.

Birondo, N. (ed.). (2022). *The Moral Psychology of Hate.* London, New York: Rowman & Littlefield Publishers.

Bloser, C., Stahl, T. (eds.) (2019). *The Moral Psychology of Hope.* London, New York: Rowman & Littlefield Publishers.

Braund, S., M., Most, G. W. (eds.) (2003). *Ancient Anger: Perspectives from Homer to Galen. Yale Classical Studies*; v. 32. Cambridge: Cambridge University Press, 2003.

Cacioppo, J., T. (1998). Somatic responses to psychological stress: The reactivity hypothesis. *Advances in Psychological Science,* 2, 87–114.

Cannon, W., B. (1927). The James-Lange theory of emotions: A critical examination and an alternative theory. *The American Journal of Psychology,* 39(1/4), 106–124.

Carminati L (2021). Emotions, Emotion Management and Emotional Intelligence in the Workplace: Healthcare Professionals' Experience in Emotionally-Charged Situations. *Frontiers in Sociology* 6, 640384. https://doi.org/10.3389/fsoc.2021.640384.

Carter, A., Gordon, E., C. (eds.) (2017). *The Moral Psychology of Pride,* London, New York: Rowman & Littlefield Publishers.

Celeghin, A., Diano, M., Bagnis, A., Viola, M., & Tamietto, M. (2017). Basic emotions in human neuroscience: Neuroimaging and beyond. *Frontiers in Psychology,* 8,1432. https://doi.org/10.3389/fpsyg.2017.01432.

Chaiken, S., Liberman, A., & Eagly, A. H. (1989). Heuristic and systematic information processing within and beyond the persuasion context. In J., S. Uleman & J., A. Bargh (eds.), *Unintended Thought* (212–252). New York: The Guilford Press.

Cherry, M., Flanagan, O. (eds.) (2017). *The Moral Psychology of Anger.* London, New York: Rowman & Littlefield Publishers.

Cokelet, B., Maley, C., J. (eds.) (2019). *The Moral Psychology of Guilt.* London, New York: Rowman & Littlefield Publishers.

Collins, D., Jovanovič, I., V., Alfano, M. (eds.) (2023). *The Moral Psychology of Trust.* London: Lexinton Books.

Conte, B., Hahnel, U. J. J., & Brosch, T. (2022). From values to emotions: Cognitive appraisal mediates the impact of core values on emotional experience. *Emotion.* Advance Online Publication. https://doi.org/10.1037/emo0001083.

Damasio, A., R. (2008) [1994]. *Descartes' Error: Emotion, Reason and the Human Brain.* New York: Random House.

Darwin, C. (2007) [1872]. *The Expression of the Emotions in Man and Animals.* New York: Filiquarian.

de Sousa, R. (1987). *The Rationality of Emotion.* Cambridge, MA: MIT Press.

Démuth, A. (2003). *Homo - anima cognoscens.* Bratislava: Iris.

Démuth, A. (2013). *Game Theory and Decision Making.* Krakow: Towarzystwo Słowaków w Polsce.

Démuth, A. (2017). *Beauty, Aesthetic Experience, and Emotional Affective States.* Berlin: Peter Lang Verlag.

Démuth, A., Démuthová, S. Keceli, Y. (2022a): A semantic analysis of the concept of beauty (güzellik) in Turkish language: Mapping the semantic domains. *Frontiers in Communication* 7(20), 1–11. https://doi.org/10.3389/fcomm.2022.797316.

Démuth, A., Démuthová, S. Keceli, Y. (2022b): A semantic analysis of the concept of anger and its connotations (based on free associations experiment) in the Turkish language. *Luboslovie.* 22, 111–127.

Denson, T., F., Pedersen, W., C., Ronquillo, J., Nandy, A., S. (2009, Apr.). The angry brain: neural correlates of anger, angry rumination, and aggressive personality. *Journal of Cognitive Neuroscience.* 21(4), 734–744. https://doi.org/10.1162/jocn.2009.21051.

Dewey, J. (1894). The theory of emotion: I: Emotional attitudes. *Psychological Review.* 1(6), 553–569. https://doi.org/10.1037/h0069054.

Eisenberg N. (2000). Emotion, regulation, and moral development. *Annual Review of Psychology,* 51, 665–697. https://doi.org/10.1146/annurev.psych.51.1.665.

Ekman, P. (1984). Expression and the nature of emotion. In Scherer, K., and Ekman, P. (eds.), *Approaches to Emotion* (319–344). Hillsdale, NJ: Lawrence Erlbaum.

Ekman, P. (1992). An argument for basic emotions. *Cognition & Emotion,* 6, 169–200. https://doi.org/10.1080/02699939208411068.

Ekman, P. (2003). Emotions inside out. 130 years after Darwin's "The Expression of the Emotions in Man and Animal". *Annals of the New York Academy of Sciences,* 1000, 1–6. https://doi.org/10.1196/annals.1280.002.

Ekman, P., & Friesen, W. V. (1971). Constants across cultures in the face and emotion. *Journal of Personality and Social Psychology,* 17(2), 124–129. https://doi.org/10.1037/h0030377.

Elpidorou, A. (ed.) (2022): *The Moral Psychology of Boredom.* London, New York: Rowman & Littlefield Publishers.

Fingerhut, J., Prinz, J., J. (2020, Apr.). Aesthetic emotions reconsidered. *The Monist,* 103(2), 223–239. https://doi.org/10.1093/monist/onz037.

Forgas, J.P. (1995). Mood and judgment: The Affect Infusion Model (AIM). *Psychological Bulletin,* 117(1), 39–66. https://doi.org/10.1037/0033-2909.117.1.39.

Frijda, N. (1988) *De emoties; een overzicht van onderzoek en theorie.* Amsterdam: Bert Bakker.

Frijda, N., H., Kuipers, P., & ter Schure, E. (1989). Relations among emotion, appraisal, and emotional action readiness. *Journal of Personality and Social Psychology,* 57(2), 212–228. https://doi.org/10.1037/0022-3514.57.2.212.

Fussi, A., Rodogno, R. (eds.) (2023). *The Moral Psychology of Shame.* London, New York: Rowman & Littlefield Publishers.

Gaulin, S., J., C., McBurney, D., H. (2003). *Evolutionary Psychology.* London: Pearson.

Goldie, P. (2000). *The Emotions: A Philosophical Exploration,* Oxford: Oxford University Press. https://doi.org/10.1093/0199253048.001.0001.

Goldie, P. (2002). Emotions, feelings and intentionality, *Phenomenology and the Cognitive Sciences,* 1(3), 235–254. https://doi.org/10.1023/A:1021306500055

Gotlib, A. (ed.) (2017). *The Moral Psychology of Sadness.* London, New York: Rowman & Littlefield Publishers.

Gotlib, A. (ed.) (2022). *The Moral Psychology of Regret,* London, New York: Rowman & Littlefield Publishers.

Gray, K., Wegner, D., M. (2011). Dimensions of Moral Emotions. *Emotion Review,* 3(3), 258–260. https://doi.org/10.1177/1754073911402388.

Griffiths P. E., Scarantino A. (2005). Emotions in the wild: The situated perspective on emotion. In Robbins P., Aydede M. (Eds.), *Cambridge handbook of situated cognition* (437–453). Cambridge, UK: Cambridge University Press.

Gu, S., Gao, M., Yan, Y., Wang, F., Tang, Y., Y., & Huang, J., H. (2018). The neural mechanism underlying cognitive and emotional processes in creativity. *Frontiers in Psychology,* 9, 1924. https://doi.org/10.3389/fpsyg.2018.01924.

Gu, S., Wang, F., Patel, N., P., Bourgeois, J., A., & Huang, J., H. (2019) A model for basic emotions using observations of behavior in Drosophila. *Frontiers in Psychology.* 10, 781. https://doi.org/10.3389/fpsyg.2019.00781

Gu, S., Wang, W., Wang, F., & Huang, J. H. (2016). Neuromodulator and emotion biomarker for stress induced mental disorders. *Neural Plasticity.* 2016, 2609128. https://doi.org/10.1155/2016/2609128.

Haidt, J. (2003). The moral emotions. In R. J. Davidson, K. R. Scherer, & H. H. Goldsmith (eds.), *Handbook of Affective Sciences* (852-870). Oxford: Oxford University Press.

Heidegger M. (1927). *Sein und Zeit*. Halle: M. Niemeyer, 1927.

Hutto, D., D., Robertson, I., and Kirchhoff, M., D. (2018). A new, better BET: rescuing and revising basic emotion theory. *Frontiers in Psychology*, 9, 1217. https://doi.org/10.3389/fpsyg.2018.01217.

Immordino-Yang, M., H., McColl, A., Damasio, H., Damasio, A. (2009, May). Neural correlates of admiration and compassion. *Proceedings of the National Academy of Sciences*, 12, 106(19), 8021-8026. https://doi.org/10.1073/pnas.0810363106.

Inan, I., Watson, L., Whitcomb, D., Yigit, S. (2018). *The Moral Psychology of Curiosity*. London, New York: Rowman & Littlefield Publishers.

Izard, C., E. (1977). *Human Emotions*. New York, NY: Plenum Press. https://doi.org/10.1007/978-1-4899-2209-0.

James, W., Lange, C.G. (1922). *The Emotions*. Baltimore: Williams & Wilkins Co.

Kawabata H, Zeki S. (2004, Apr.). Neural correlates of beauty. *Journal of Neurophysiology*, 91(4), 1699-1705. https://doi.org/10.1152/jn.00696.2003.

Keltner, D., & Haidt, J. (2003). Approaching awe, a moral, spiritual, and aesthetic emotion. *Cognition and Emotion*, 17(2), 297-314. https://doi.org/10.1080/02699930302297.

Kluemper, D., H., DeGroot, T., & Choi, S. (2013). Emotion management ability: Predicting task performance, citizenship, and deviance. *Journal of Management*. 39, 878-905. https://doi:10.1177/0149206311407326.

Kvajo, M. (2016). What we talk about when we talk about emotions. *Cell*, 167, 1443-1445. https://doi.org/10.1016/j.cell.2016.11.029.

Laurence, S., Margolis, E. (2003). Concepts and conceptual analysis. *Philosophy and Phenomenological Research*, 67 (2), 253-282.

Lazarus, R., S., Folkman, S. (1984). *Stress, Appraisal, and Coping*. New York: Springer Pub.

LeDoux, J., E. (1996) *The Emotional Brain*. New York: Simon and Schuster.

Mason, M. (ed.) (2018). *The Moral Psychology of Contempt*. London, New York: Rowman & Littlefield Publishers.

McIntosh, D., N., Zajonc, R., B., Vig, P., B., Emerick, S., W. (1997). Facial movement, breathing, temperature, and affect: Implications of the vascular theory of emotional efference. *Cognition & Emotion*, 11(2), 171-195. https://doi.org/10.1080/026999397379980.

Menninghaus, W., Schindler, I., Wagner, V., Wassiliwizky, E., Hanich, J., Jacobsen, T., & Koelsch, S. (2020). Aesthetic emotions are a key factor in aesthetic evaluation: Reply to Skov and Nadal (2020). *Psychological Review*, 127(4), 650-654. https://doi.org/10.1037/rev0000213.

Menninghaus, W., Wagner, V., Wassiliwizky, E., Schindler, I., Hanich, J., Jacobsen, T., & Koelsch, S. (2019). What are aesthetic emotions? *Psychological Review*, 126(2), 171-195. https://doi.org/10.1037/rev0000135

Ortony, A., Turner, T. J. (1990). What's basic about basic emotions? *Psychological Review*, 97, 315-331.

Panksepp, J. (2007). Neurologizing the psychology of affects: How appraisal-based constructivism and basic emotion theory can coexist. *Perspectives on Psychological Science*, 2, 281-296. https://doi.org/10.1111/j.1745-6916.2007.00045.x.

Perlovsky L (2014) Aesthetic emotions, what are their cognitive functions? *Frontiers in Psychology*. 5(98). https://doi.org/10.3389/fpsyg.2014.00098

Pismeny, A., Brogaard, B. (eds.) (2022). *The Moral Psychology of Love*. London, New York: Rowman & Littlefield Publishers.

Pizarro, D., A. (2007). Moral emotions. In Baumeister, R., F; Vohs, K., D. (eds.). *Encyclopedia of Social Psychology* (588-589). SAGE Publications, Inc. https://doi.org/10.4135/9781412956253.n350.

Ploder, A., Eder, A. (2015). Semantic differential, In Wright, J., D. (eds.). *International Encyclopedia of the Social & Behavioral Sciences (Second Edition)* (563-571), Amsterdam: Elsevier.

Prinz, J., (2009). The moral emotions. In Goldie, P. (ed.), *The Oxford Handbook of Philosophy of Emotion* (2009; online edn, Oxford Academic, 2 Jan. 2010), https://doi.org/10.1093/oxfordhb/9780199235018.003.0024.

Prinz, Jesse J. (2004). *Gut Reactions: A Perceptual Theory of Emotion*. Oxford: Oxford University Press.

Protasi, S. (ed.) (2022). *The Moral Psychology of Envy*. London, New York: Rowman & Littlefield Publishers.

Ratcliffe, M. (2005). The feeling of being. *Journal of Consciousness Studies*, 12(8-10), 43-60.

Reber, A., S. (1995). *The Penguin Dictionary of Psychology*. London: Penguin Books.

Roberts, R., Telech, D. (eds.) (2019). *The Moral Psychology of Gratitude*. London, New York: Rowman & Littlefield Publishers.

Robinson, B. (ed.) (2021). *The Moral Psychology of Amusement*. London, New York: Rowman & Littlefield Publishers.

Rudolph, U., & Tscharaktschiew, N. (2014). An attributional analysis of moral emotions: naïve scientists and everyday judges. *Emotion Review*, 6(4), 344-352. https://doi.org/10.1177/1754073914534507.

Russell, J. (2003). Core affect and the psychological construction of emotion. *Psychological Review*, 110, 145-172. https://doi.org/10.1037/0033-295X.110.1.145.

Russell, J. A. (2006). Emotions are not modules. *Canadian Journal of Philosophy*, 32, 53-71. https://doi:10.1353/cjp.2007.0037.

Saarimaki, H., Gotsopoulos, A., Jaaskelainen, I. P., Lampinen, J., Vuilleumier, P., Hari, R., et al. (2016). Discrete neural signatures of basic emotions. *Cerebral Cortex*, 26, 2563-2573. https://doi.org/10.1093/cercor/bhv086.

Sartwell, C. (2004) *Six Names of Beauty*. New York: Routledge.

Scarantino, A., de Sousa, R. (Summer 2021 Edition). "Emotion". In Edward N. Zalta (ed.), *The Stanford Encyclopedia of Philosophy*. URL = <https://plato.stanford.edu/archives/sum2021/entries/emotion/>.

Scarantino, A., and Griffiths, P. (2011). Don't give up on basic emotions. *Emotion Review*. 3, 444–454. https://doi.org/10.1177/1754073911410745.

Schachter, S., Singer, J. (1962). Cognitive, social, and physiological determinants of emotional state. *Psychological Review*. 69(5), 379–399. https://doi.org/10.1037/h0046234.

Scheler, M., F. (1957). Über Scham und Schamgefühl. In Scheler, M. (ed.). *Schriften aus dem Nachlaß*, (63-154) Francke: Band I: Zur Ethik und Erkenntnislehre, zweite, durchgesehene und erweiterte Auflage. Bern.

Scherer, K., R., & Coutinho, E. (2013). How music creates emotion: A multifactorial process approach. In T. Cochrane, B. Fantini, & K. R. Scherer (eds.), *The Emotional Power Of Music: Multidisciplinary Perspectives on Musical Arousal, Expression, and Social Control* (121–145). Oxford: Oxford University Press. https://doi.org/10.1093/acprof:oso/9780199654888.003.0010.

Schindler, I., Hosoya, G., Menninghaus, W., Beermann, U., Wagner, V., Eid, M., et al. (2017) Measuring aesthetic emotions: A review of the literature and a new assessment tool. *PLoS ONE* 12(6): e0178899. https://doi.org/10.1371/journal.pone.0178899.

Seo, M.-G., Bartunek, J. M., & Feldman Barrett, L. (2010). The role of affective experience in work motivation: test of a conceptual model. *Journal of Organizational Behavior*. 31, 951–968. https://doi.org/10.1002/job.655.

Seok, B. (2017). *Moral Psychology of Confucian Shame. Shame of Shamelessness*. London, New York: Rowman & Littlefield Publishers.

Silvia, P. (2009, February). Looking past pleasure: Anger, confusion, disgust, pride, surprise, and other unusual aesthetic emotions. *Psychology of Aesthetics, Creativity, and the Arts*, 3(1), 48–51. doi:10.1037/a0014632.

Silvia, P. J. (2005). What is interesting? Exploring the appraisal structure of interest. *Emotion*, 5(1), 89–102. https://doi.org/10.1037/1528-3542.5.1.89.

Skov, M., Nadal, M. (2020). There are no aesthetic emotions: Comment on Menninghaus et al. (2019). *Psychological Review*, 127(4), 640–649. https://doi.org/10.1037/rev0000187.

Solomon, R., C. (1993). *The Passions: Emotions and the Meaning of Life*. Indianapolis: Hackett Publishing.

Song, Y., and Hakoda, Y. (2018). Selective impairment of basic emotion recognition in people with autism: discrimination thresholds for recognition of facial expressions of varying intensities. *Journal of Autism and Developmental Disorders*. 48, 1886–1894. https://doi.org/10.1007/s10803-017-3428-2.

Stallen, M., Rossi, F., Heijne, A., Smidts, A., De Dreu, C., K., W., Sanfey, A., G. (2018, Mar). Neurobiological mechanisms of responding to injustice. *Journal of Neurosciience*, 21;38(12), 2944–2954. https://doi:10.1523/JNEUROSCI.1242-17.2018.

Steinbock, A., J. (2014). *Moral Emotions: Reclaiming the Evidence of the Heart.* Evanstone: Northwestern University Press.

Strohminger, N., Kumar, V. (eds.) (2018). *The Moral Psychology of Disgust.* London, New York: Rowman & Littlefield Publishers.

Taleb, N., N. (2001) *Fooled by Randomness: The Hidden Role of Chance in the Markets and in Life.* New York: Texere.

Tangney, J. P., Stuewig, J., & Mashek, D. J. (2007). Moral emotions and moral behavior. *Annual Review of Psychology,* 58, 345–372. https://doi.org/10.1146/annurev.psych.56.091103.070145.

Tangney, J., P., Stuewig, J., Mashek, D., J. (January 2007). Moral emotions and moral behavior. *Annual Review of Psychology,* 58(1), 345–372.

Tomkins, S. (1962). *Affect Imagery Consciousness: The Positive Affects. Vol. 1.* New York: Springer Publishing.

Vetter, N. C., Drauschke, M., Thieme, J., & Altgassen, M. (2018). Adolescent basic facial emotion recognition is not influenced by puberty or own-age bias. *Frontiers in Psychology.* 9, 956. https://doi.org/10.3389/fpsyg.2018.00956.

Von Scheve, C. (2012). Emotion regulation and emotion work: Two sides of the same coin? *Frontiers in Psychology,* 3, 496. https://doi.org/10.3389/fpsyg.2012.00496.

Wang, F., Pan, F., Shapiro, L. A., & Huang, J. H. (2018). Stress induced neuroplasticity and mental disorders 2018. *Neural Plasticity,* 2018, 5382537. https://doi.org/10.1155/2018/5382537.

Weiss, H., M., Beal, D., B. (June 2005). Reflections on affective events theory. Emotion. *Research on Emotion in Organizations,* 1, 1–21. https://doi.org/10.1016/S1746-9791(05)01101-6.

Williams, R. (2017). Anger as a basic emotion and its role in personality building and pathological growth: the neuroscientific, developmental and clinical perspectives. *Frontiers in Psychology,* 8, 1950. https://doi.org/10.3389/fpsyg.2017.01950.

Wilson-Mendenhall, C., D., Barrett, L., F., & Barsalou, L., W. (2013). Neural evidence that human emotions share core affective properties. *Psychological Science,* 24, 947–956. https://doi.org/10.1177/0956797612464242.

Wittgenstein, L. (1921). *Logisch-Philosophische Abhandlung [Tractatus Logico-Philosophicus],* In Wilhelm Ostwald (ed.), *Annalen der Naturphilosophie,* Vol. XIV, issue 3/4.

On Some Etymological, Grammatical and Contextual Reasons for the Vagueness of the Concept of Beauty[1]

Andrej Démuth, Slávka Démuthová, Yasin Keceli

---
[1] The text is the original version of the study, which in a modified and translated version was published in Turkish as Keceli, Démuth, Démuthová 2021.

Abstract. This chapter aims to thematise selected semantic (etymological, grammatical and contextual) reasons for the vagueness of the concept of beauty. The chapter is the result of analyses of concepts and basic terms, which are used to denote the concept of beauty in various (especially Indo-European) languages, their etymological connections, semantic dimensions and contextual personalities. The authors aim to show that the concept of beauty belongs to the overarching themes, and its vagueness lies both in the multidimensionality of the given concept in each language (it denotes different levels of understanding – object, intentional content, feeling) and in the different semantic dimensions by which the given concept is denoted in selected specific languages (different aspects of the concept in different languages). The text forms part of a wider project of mapping, cognitive-semantic analysis and the creation of semantically "purer" concept of beauty – especially the use of Gärdenfors' model of semantic spaces for mapping the conceptual domains of aesthetic concepts.

Introduction

In 2004, Crispin Sartwell wrote the book *Six Names of Beauty*, in which he showed that beauty in different languages is not only denoted by the same terms, but the terms we use to denote it denote significantly different contents (Beauty as an object of longing), Hebrew (Yapha as glow, bloom), Sanskrit (Sundara – holiness), Greek (to kalon – idea), Japanese (Wabi-sabi – perfection in imperfection) or in Navajo (Hozho – health, harmony – Sartwell, 2004). Is there even a unified concept of beauty? Or are they just different names for the same thing? What do we mean by the term "beauty"? Is it an idea, a thing, the properties of things, or is it our feelings, an intentional object or something different?

There is no doubt that beauty is one of the fundamental concepts in philosophy and the study of aesthetics. Some philosophers believe that it was particularly beauty that led the Pythagoreans to their interest in the world and the study of numbers and ratios (Vopěnka, 2001, 52–54) and also that it is particularly in beauty that the truth, wisdom and goodness are joined together; therefore, it is beauty that should be the ultimate goal of any philosophical study (Schelling, 1800/1907, 631).

Despite (sometimes more, sometimes less) intense study that has already lasted more than two and a half millennia, we still do not have a satisfactory and generally accepted definition of what beauty is and what constitutes it. Many scholars, in fact, seem to hold the opinion that beauty is in its essence indefinable, that there are an infinite number of types of beauty (Levinson, 2011) and diametrically different elements, aspects or components thereof. Thus it is not possible for us to wholly and sufficiently comprehend it. That is why scholars often focus on the study of its accompanying notions (attractiveness, the sublime, elegance, grace, etc.) and its internal diversification – on the study of its elements, aspects and related factors.

The subject matter of this study is an analysis of the reasons why the notion of beauty is one of the vaguest notions, and not only in the study of aesthetics. One of the possible reasons that beauty is problematic and difficult to grasp is its potential multidimensionality. Beauty is, so to speak, saturated from many (different and often even contrary) areas and dimensions and therefore it is often very difficult to wholly gather them under a single umbrella notion. Such a multidimensional understanding of the notion of beauty also accounts for the use of the concepts in everyday language.

Etymological Aspects

If we look at the words used to designate beauty in various languages, we will see that the most basic words used to denote this concept are often derived from wholly different terms or etymological roots. In English, for example, the concept is linked to the expression "beauty" that has its roots in Old French and Latin. The original French expression for beauty that gave rise to the English term "beauty", is "bel", and that is linked with the Latin "bellus". According to Etymonline.com it comes from the "early 14c., *bealte*, 'physical attractiveness', also 'goodness, courtesy', from Anglo-French *beute*, Old French *biauté* 'beauty, seductiveness, beautiful person' (12c., Modern French *beauté*), earlier *beltet*, from Vulgar

Latin *bellitatem* (nominative *bellitas*) 'state of being pleasing to the senses' (source also of the Spanish *beldad*, Italian *belta*), from Latin *bellus*" (Harper, 2020). The *Cambridge Advanced Learner's Dictionary* describes this concept as "the quality of being pleasing, especially to look at, or someone or something that gives great pleasure, especially when you look at it" (McIntosh, 2013, 125). The chief meaning of the English concept is therefore connected with the denotation of a pleasant sensory stimulus. On the other hand, the *Oxford Advanced Learner's Dictionary* shifts the definition of this concept by defining beauty as "the quality of giving pleasure to the senses or to the mind" (Hornby, 2011, 119). According to this definition, beauty is not merely a matter of sensory perceptions (is not merely in the eye (or ear, ...) of the beholder – beauty is not located only in the senses and (figuratively) only on the surface of the skin), but it is also an object of rational or mental pleasure. We do not have to perceive beautiful things with our senses alone but can imagine them with the help of our imagination or memory. In both cases the "beautiful" objects are something we like – an image of what we feel is pleasant – something that evokes feelings of pleasure.

The situation is similar in other languages that are based on Latin terminology (the romance languages: Spanish, Italian, etc.). The etymological roots and original meaning of this term that comes from the Latin "bellus" are based on a certain (sensory or rational) hedonistic attitude. A beautiful object is something that is "visible" (to our senses or the "inner eye") and the perception of such an image evokes in us a reflection of our own pleasant state which arises through the perception of the particular objects.

We can find the same sensory "logic" of beauty in Turkish, German or Swedish. According to Nisanyan (2011, 308), the Turkish word for beauty, "güzel", comes from Old Turkish and developed from the word "gözel", which in Central Turkey means "a pleasant form, beautiful". This word comes from the Old Turkish "köz", "göz". The Swedish or Proto-Germanic word "vacker" ("wakraz" – Kluge, 1891) refers to the sweet pleasure that arises from the perception of an object. The German expression "die Schönheit" ("skönhet" in Swedish) points to the "visibility" (old German: Scōni "ansehnlich") and "appearance" (schauen). The authors of the dictionary *Der deutsche Wortschatz von 1600 bis heute* (DWDS) assume that the original meaning of this word was "visible, alive", from which "beautiful" and "good, pleasant" (also regarding auditory perceptions) have developed. The root of this word points to the fact that beauty not only relates to the ability to see but also to give attention, observe. Simply, the beauty of objects or that in our imagination draws our attention and lets us see that which is worthy of attention. The adjective "schön" therefore is also derived from the meaning of the word "Schönheit", which directly refers to the objects themselves (such as decorations and jewellery – not to the perception of them) and to their exceptional properties (Pfeifer, 1993). We therefore use the concept of beauty to not only denote the pleasantness of what we see in our imagination but also to ontologise that which provokes these notions and sensations. We cross the border from perception directly to the object. In Latin, besides the term "bellus", we also use the terms "pulchritūdō" (probably from *polcher*, the spelling with -ch- is Hellenised, hence we may assume the root *polkro- or *pelkro-. Walde and Hoffmann (Walde, Hoffmann, 1938), and Pokorny (Pokorny, 1959) assume that this belongs to the PIE *perkr (colourful)) or "formōsitās", "venustās", which come from other roots (form, Venus). But they all denote a particular aesthetic uniqueness, excellence or formal perfection rather than sensory pleasantness (Špaňár, Hrabovský, 1998, 492, 806). Similarly, several terms to denote beauty are available in Greek, the most fundamental of which is "κάλλος". Kallos firstly denotes the beauty of the body and secondarily the beauty of a particular individual – as a rule it is exclusively used to describe women and teenage girls (s. a. 2015). (It is interesting to note that the Latin expression "bellus", similar to the English "beauty", tended to be used almost exclusively to describe women and children, and it was only possible to use it to describe a man ironically, belligerently or offensively – the word that applies to men is "handsome".) This understanding was only changed by Shakespeare's use of the notion of "beautiful" in his *Sonnets*, where it is apparent that the first 126 are dedicated to the beauty of a young male aristocrat and another 25 to the beauty of a dark-haired lady. In ancient times the way of thinking about the attractiveness of a man was understood in the hidden and spiritual sense (Cunliffe, 1924). The Neo-Platonic Marsilio Ficino spoke about it as an expression of the Platonic "cosmic love", which could not be publicly associated with sexuality: that was reserved for the vulgar – the biological – love for a woman. For the understanding of beauty and sexuality in Elizabethan English, see e.g., Hilský, 1997. Only later is it possible to use the notion of kallos to describe things such as clothes and jewellery, i.e., things that make their wearer more beautiful (Liddell & Scott, 1940). German (though not only German) even has a word to specifically express this addition of beauty or the highlighting of beauty ("schönen"). It is possible to add beauty to things – to supply them with beauty (in the same way that it was possible to lend the girdle of Aphrodite to the Graces – the Charites). It is possible to decorate and embellish things, but it is also possible to improve them, cover up their deficiencies and improve their appearance, form or features – through direct intervention in the things themselves (e.g., "Färbungen verbessern, Wein klären"), not only in our perception of them.

A similar ontologising understanding of beauty may be found in some Slavic languages. In Russian, for example, the notion of beauty uses terms derived from the Old Slavic denotation of the colour "red" (краса/красота / *krasьnъ – Derksen, 2008, 246). The Proto-Slavic *krasa was linked with *shine, red, colour of the fire* and with the Proto-Slavic *kresati* (Rejzek, 2001, 310). The *Concise Etymological Dictionary of the Slovak Language* provides an association with fire and shine (the term also occurs in cremation burial sites – Králik, 2015, 297), but especially with decoration, adornment and embellishment. To embellish (adorn) here means to add shine, or specifically to paint something red, a colour attractive at that time (or later with any other colour) to adapt the surface (e.g., of "kraslica", a painted Easter egg). From this we can infer two different meanings of embellishment: (1) an improvement of the object's true qualities in such way that they evoke a greater feeling of liking; (2) beautifying/polishing – pretending that an object has qualities that it lacks in reality, with the intention of improving the impression it makes ("a red blush on the face is the colour of health, good looks, beauty" – Králik, 2015, 297).

A different aspect of the notion of beauty may be found in Polish terminology. The basic term for beauty in Polish is "piękny". This is etymologically identical to the Czech "pěkný" and Slovak "pekný" and comes from the Proto-Slavic "pěkrъ" (its root is similar to the Latin pulcher) or from the Old Slavic пѣгъ. The *Concise Etymologi-*

cal Dictionary of the Slovak Language shows that this might be a derivation of the IE *pej(e) (to be fat, to abound in something, to exuberate – cf. the Old Indian *páyate* (it heaves, it exuberates), but also the Avestan root (to have milk in one's breasts – Králik, 2015, 431). Strangely, however, while in Czech and Slovak "pekný" refers to a lesser degree of attractiveness, the Polish term is the second level of irregular comparison. The Polish equivalent of this word is rather the term "ładny" (from the Proto-Slavic ladъ – Żmigrodzki, 2018). The notion "ładny" refers to good looks, but in a different context, as well as to tidiness and harmony – harmony that gives rise to delectability – to be pleasing to the eye. We like to experience pleasing, graceful movements not only because they are easy on the eyes but also because their suitability comes from their fluency, they are natural and we are able to observe and fluently predict them without difficulty. Graceful movements are regulated; they abide by certain rules, patterns and order; and that makes them predictable. They are appropriate to the context, they are apt and natural – and abide by a certain order. That is why the Hebrew word "hpy", transliterated "Yapheh", denoting beauty in the Biblical Hebrew, particularly expresses appropriateness, regularity and order (Clark, 1999). Appropriateness, consonance and harmony express one of the most important aspects of beauty. It is not only because a beautiful thing is proportionate, appropriate and somehow internally harmonised with its surroundings (the appropriateness of the object and the context) but above all because beauty is something that resonates with us, with our needs and preferences (the appropriateness of the object and the observer). After all, if we like something, then it resonates – it harmonises with us – with our taste, it moves us and provokes a positive response (Démuth, 2018, 143–159).

On the other hand, in a number of languages (e.g., Semitic) it is not possible to clearly determine the external etymology of the term "beauty" and the notion it refers to may rather be found in the context and the most frequent connotation of this term. We usually translate the Arabic term "al-Jamal" as "beauty", but its understanding in the pre-Islamic period differs from the understanding of beauty in the Quran, where it is undoubtedly and especially connected with the omnipotence and perfection of God (Vilchez, 2017, 32). In a similar way, we understand the Chinese term "Měilì" as "beautiful", but depending on the whole expression it may refer to physical attraction ("měirén" – beautiful woman), a pleasant taste ("měishí" – good taste), as well as improvement ("měiróng"), good, pleasure or wish ... (Chen, Qi and Hao, 2018).

Therefore, we run up against another serious complication, that of the interchange of beauty with good, or perfection. Is this the result of etymological and semantic relations or merely historical connotations, which for some reason exist between them? The original French expression for beauty, from which the English term "beauty" comes, is "bel", which is linked to the Latin "bellus". The history of this Latin root is very old and also rather complicated. The term "bellus" in Old Latin is often connected with the word "bonus" (good), or even with "optimus" (the best). Both "bonus" and "optimus" originally derive from the Old Latin version of the word "duonus", which means "good". The Proto-Italic root of this word is *dwenos and from this root is also ultimately derived the Proto-Indo-European *dāu- and *dw-ene- (after de Vaan 2008, 73–74 – the Latin bene <abl. sg. *duened> documents that the change of *duo was caused by the non-front vowel in the following syllable. "Bellus" reflects *duenelos> *duenlos> *duelhs * *bellos). Hence the link between the "beautiful", the "good" or even the "divine" is no accident.

In Slovak we usually differentiate between "pekný" (beautiful) and "dobrý" (good), even though they might have the same etymological origin. The *Concise Etymological Dictionary of the Slovak Language* defines the adjective "lepší" (better) as the comparative of "lepý" (today merely a poeticism of the meaning of "beautiful") with a reference to "veľko-lepý" (spectacular) and the originally Indo-European *leip* (to baste with fat, anoint – Králik, 2015, 323). However, in everyday language, especially when influenced by Immanuel Kant's philosophy, we perceive a difference between the aesthetic and the ethical, between liking and a moral or otherwise axiological evaluation. However, the differentiation of these two qualities is from a certain perspective, in particular methodological, and it is semantically not as clear as it might seem at first sight. Often we identify the "beautiful" with the "good", and likewise, the "good" with the "beautiful" (Schiller's ideal of a beautiful person assumes the necessity of moral beauty – goodness – Schiller, 2005). The reasons for this interchangeability do not merely lie in the metaphorical nature of notions and our inconsistent use of language but also in the fact that "beautiful" stimuli evoke feelings very similar to those that arise from stimuli that we consider to be "good". Thanks to the neural correlates of our reward system, we "like" the "good" things (behaviour that we assess as "good"), and thanks to their pleasantness, we assess "beautiful" things as "good" (Démuth, 2019). That which connects both the notions and terms is therefore the pleasurable feelings they evoke and the fact that we perceive them as something we like, but also that for evolutionary reasons we perceive the "beautiful" as "good" for posterity (Démuth, 2019).

However, we also interchange the notion of "beautiful" with the notion of "good" in another meaning, as an expression of the level of elaboration. A "beautifully" painted picture is often an expression used to describe "well" painted picture. That is also true when the object depicted in the painting is ugly but is painted "well" – it is alive, apt, persuasive – it is "beautiful". Therefore, here beauty is linked with good in the sense of perfection, the quality of elaboration, precision, etc. In this sense we use the term "szép" in Hungarian. "Szép" refers to skilfulness and experience. The Serbo-Croatian "lep" or "lijep" opens the question of the relationship between beauty and attractiveness (after Derksen 2008). The etymology of the Serbo-Croatian term "lepota" may be derived from *lěpъ (glue → paint → rich, high-quality → beautiful) as good, as well as the quality of beauty and its linguistic comparison (better).

Grammatical Specifics

In most languages the basic terms for beauty are mainly adjectives. As for most adjectives and also for the adjective "beautiful", we intuitively presuppose the existence of three basic levels (forms) of the particular quality. It is possible to intensify it, by a gradation from the positive (1st degree – "krásny" (beautiful)) to the comparative (2nd degree – "krajší" (more beautiful)) to the 3rd degree – superlative. This intensification expresses the relative (comparative) relationship between the individual expressions. The expression "najkrajší" (the most beautiful) as a relative superlative always expresses a relationship with its surroundings (it refers to something that is "less beautiful"). In Latin we can express this relationship with the preposition "di" (from) or the conjunction "quod" (which) similarly as we do in the comparative form. In Slovak, however, the word "krásny" (beautiful) can also be formed to express a high degree of quality without any relative relationship, that is comparison. The elative, as we call this degree, is formed with the help of prefixes – "prekrásny", "čarokrásny", "divukrásny", expressing a certain absolute degree of extent. The absoluteness of this superlative thus consists in the fact that it is not a comparative expression. Rather we express it through the addition of the prefix pre- or the adverb "veľmi" (very), "mimoriadne" (exceptionally), "nesmierne" (greatly), ... (in Latin with the suffix -ssimo/a/i/e). In Slovak as well as a number of other languages we also see another – the highest degree of the relative superlative, that being in the sense of the expression of the absolute – the fully saturated expression of quality. That may also be expressed through the pronoun "samý" and the superlative "najkrajší", by which we express the absolute degree of beauty – "samý najkrajší". This expression ("the very most beautiful" in English) denotes the maximum possible degree of saturation of that particular quality and its extreme – the upper border of what is possible. The "samý najkrajší" is one and only, and it has no equal in the world (in the group, etc.).

In this context, the so-called equative presents a very specific expression of a relative comparison of beauty. It expresses equality of degree. We could then translate the Welsh expression "mwy swynol" (swynol: mor swynol: mwy swynol: mwyaf swynol – beautiful: more beautiful: equally beautiful: the most beautiful) as "equally beautiful" (Čermák, 2011, 159). However, in most languages such an independent and terminological form does not exist.

It is also possible to form the second and the third degree (elative) of the adjective "krásny" in a descriptive way. Here we can use an expression that consists of the basic form of the adjective with the second and third-degree adverbs more, very, etc. In the case of the adjective "krásny", this process is rather rare in Slovak. On the contrary, in English this is a standard form of intensification of this adjective (beautiful, more beautiful, most beautiful).[2] In Greek and some other languages the comparative and superlative forms are identical and only differ by the article put before the adjective. In other languages (e.g., English, German), the use of the definite and non-definite article enables us to distinguish between the relative and absolute forms of the superlative.

In many languages we find that the superlative form of the adjective is replaced by lexical suppletion (e.g., "veľkolepý" instead of "najkrajší" – in English "gorgeous", German "herrlich", French "magnifique") which further complicates the whole issue. The point is that here the adjective "beautiful" is replaced by a different dimension of the semantic space and its denotation. Thus, we learn that something is "strašne krásny" (awfully beautiful). By this we do not mean that sensations that are experienced are terrible and they hurt us but rather that they are extremely beautiful and pleasant. On the contrary, there may be a form of fear that we like, that would be "krásne strašný" (beautifully awful).

An oddity of language is that although we can make a regular comparison of the word "pekný" (pekný, peknejší, najpeknejší) just as we can with "krásny" (krásny, krajší, najkrajší), we normally do not consider the expression "krajší" to be the comparative form of the "krásny" but rather as the (irregular) comparative form of the adjective "pekný". As if the comparison of "pekný" automatically presupposed a change of essence – of quality from "pekný" to "krajší" rather than a change in intensity or rate (through the comparative) and the preservation of the same essence ("peknejší"). The interesting thing is that in Slovak this process may only be applied to five adjectives (dobrý (good), zlý (bad), veľký (big), malý (small), pekný (beautiful)) and only the opposite of the term "pekný" – "škaredý" – does not have an analogous irregular comparative form.

The relationship of the adjective "krásny" to the noun "krása" is similarly unclear. In many languages it is obvious that the noun was derived from the original adjective (e.g., German "Schönheit" undoubtedly indicates that the adjective "schön" came first and then the noun was created using the suffix "-heit"). In English we have the opposite relationship and we see that "beauty" came first, "beautiful" being a quality derived from saturation with beauty. It

---

2 French: beau, plus beau, plus beau.

is then questionable whether beauty is something that truly exists as a property of a percept or item, with the various degrees of saturation, or whether it is merely a concept that we have derived from a comparison of the individual impulses and their evaluation, subsequently creating a universal notion to represent this quality.

Contextual Specifics

A separate aspect of the use of the notion of beauty is its contextual (especially historical and social) facets. The Arabic expression "jamal" represents a good example of a historical influence on the understanding and meaning of notions. In addition to this expression representing a masculine noun and at the same time a form of masculine name (as opposed to the usual perception of beauty as a feminine noun), its meaning has essentially changed throughout history depending on the historical and religious context. Whereas in the pre-Islamic period beauty was especially perceived as the source of sensual pleasure and joy, in the Islamic period it was rather spiritual; it was proof of the goodness and omnipotence of God – Allah. The Quran understands physical and sensory beauty as being less valuable; perfection, control, dignity and morality became the central domain of beauty in the Islamic understanding. We can find similar historical shifts in European cultural development. Umberto Eco (Eco, 2005) notes that beauty, its understanding, connotations and ideals frequently changed depending on the varying geographic-cultural-historical contexts and conditions. Beauty was often represented as an expression of proportion, regularity and harmony. The appropriateness and proportionality of things often led us to a transcendental or even magical perception of beauty. We believed that beauty is organised and that its study would lead us to a deeper order and to the meaning of the order of things. At other times, our understanding of beauty specifically preferred that which disrupts this organisation – spontaneity, unrest and incomprehensibility. That it is impossible to comprehend, understand and analyse beauty provokes in us an intellectual interest, and its scope which cannot be grasped through our senses produces a category of "the sublime". The opposing position is often taken by simplicity and elegance.

Japanese aesthetics recognises a number of aesthetic principles and ideals that fulfil the idea of beauty (Parkes, Loughnane, 2018). Some of them are based on imperfection (the ideal of Wabi-sabi – especially related to asymmetry, irregularity (Fukinsei), simplicity (Kanso), the acknowledgement of history (Koko), naturalness (Shizen), gentleness and a lack of pomposity (Yūgen), freedom and spontaneity (Datsuzoku), tranquility (Seijaku)), but they also studied perfection (Miyabi), a delicate simple beauty (Shibumi), personal grace (Iki), depth of meaning (Yūgen) or even the acknowledgement and accentuation of existing damage and the way it is overcome (Kintsugi).

Similarly, in other, often distant cultures we can not only find different ideals of beauty (ergo that which represents the specific ideal of beauty – e.g. scarred skin, tattoos, a long neck, small feet, a stretched lower lip or accentuating it with, e.g., botox, etc.) but also a different understanding of what actually constitutes beauty and its chief semantic dimensions. Therefore, it is understandable that at various periods, in various cultures and societies different aspects of beauty are preferred and emphasised. What is interesting is that we can only understand many of these aspects when we begin to "read" and comprehend their evolutionary, historical or cultural context, including the fact that these culturally conditioned aspects of beauty are translatable and even transformable into other cultural and historical contexts – this is documented by the various fashion trends and ideals of aesthetic feeling (e.g., pale skin in Japan, changes to the shape of the eyes in Korea and so on), and in these contexts they may be considered to be attractive. Although not all the objective characteristics of aesthetic objects always necessarily produce feelings of liking, it seems that almost everywhere in the world we find concepts that characterise the semantic space we understand as "beauty", denoting very similar semantic contents. These may not be due to merely the objective characteristics of objects, and in various languages these semantic spaces may not even be divided in the same way. But it does seem that our conceptual schemes that feed the notion of beauty are based on universal or very similar sensations and aesthetic experience that we have when we perceive something beautiful, although naming some of the dimensions of this semantic space (of the notion of beauty) is undoubtedly culturally and historically conditioned.

Conclusion

From this etymological, grammatical and contextual analysis, it follows that the semantic analysis of the notion of beauty represents a very complicated task that needs to take place on several levels. It seems that although we intuitively understand the meaning of the concept of beauty and some of its chief semantic dimensions, as soon as we attempt to comprehend it in a more precise and semantically clear way, the detailed and complete structure of this notion often wholly escapes us. In this study we have attempted to point out that one of the reasons for problems in comprehending the meaning of the notion of beauty is that it is an abstract overarching concept that draws on a number of diverse areas. Inter alia, that is manifested in the fact that in various languages, the basic term that denotes the concept of beauty is derived from wholly different roots and bases. We have shown that one of the most important dimensions of this concept is the designation of pleasant and less pleasant feelings which are brought about by the aesthetic experience and aesthetic evaluation. Among other dimensions frequently used to designate the notion of beauty belongs perfection – the quality of elaboration, goodness, harmony, exceptionality and some others. This correlates with our previous theoretical research – the conceptual analysis of the notion of beauty, according to which, among the chief dimensions of beauty in addition to pleasantness, belong subjectivity and objectivity, perfection-imperfection, exceptionality-commonplaceness, simplicity-complexity and activity-passivity (Démuth 2017, 2020; Démuth, Démuthová, 2018). Many of these dimensions are directly contained in the basic aesthetic terms by which we denote the concept of beauty in various languages. However, Hosoya and Menninghaus (Hosoya et al., 2017) have noted, in their research, that while in a number of more specific and concrete aesthetic categories there is a relatively high internal cohesion between the use of aesthetic notions and categories (for instance beauty, elegance, grace, and sexiness – Menninghaus et al., 2019) by scholarly and non-scholarly users of the natural language, the lowest internal correspondence between all users of the language is in the most general and overarching notion of beauty. It means that this concept draws on the greatest number of dimensions and therefore its definition is rather problematic, which is often manifested in the so-called subjective nature of aesthetic judgement. That, however, may not be an issue of a lack of objectivity (and therefore in the difference of reference frameworks – contexts in which we make the assessment) but rather in the users of the language who focus on different fundamental dimensions of the notion and thus on different semantic areas to those which the particular user (and the notion selected by them) refers.

This brings us to the conclusion that a closer understanding of the concept of beauty requires a detailed analysis of more specific aesthetic notions and dimensions which saturate this concept in various cultural-linguistic contexts, both through empirical analyses of the fundamental aesthetic concepts in various users of language (an approach using empirical aesthetics and linguistics with a top-down method) as well as through an analysis of their mutual semantic relationships. That seems to be a way that we might remove the lack of clarity and semantic vagueness of the notion of beauty in various languages but to also understand more closely how the particular cultural-historical-linguistic contexts influence our manner of thinking about things, as well as the manner of the subsequent classification of the world and aesthetic experience in the particular linguistic environment.

References

Clark, M. (1999). *Etymological Dictionary of Biblical Hebrew: Based on the Commentaries of Samson Raphael Hirsch*. Jerusalem: Philipp Feldheim.

Cunliffe, R., J. (1924). *A Lexicon of the Homeric Dialect*. Oxford: Oxford University Press.

Čermák, F. (2011). *Jazyk a jazykověda*. [Language and Linguistics] Praha: Karolinum Press.

Démuth, A. (2017). Conceptual Analysis of the Concept of Beauty in Cognitive-Scientific Research. In: Démuth, A. (ed.). *The Cognitive Aspects of Aesthetic Experience – Introduction*. Frankfurt am Main: Peter Lang, 31–52.

Démuth, A. (2018). Heidegger a problém (ne)naladenosti. [Heidegger and the problem of (dis)attunement]. In: Leško, V., Mayerová, K. (eds.). *Heidegger v Česku, Poľsku a na Slovensku*. Košice: Univerzita Pavla Jozefa Šafárika v Košiciach, 143–159.

Démuth, A., Démuthová, S. (2018). The Problem of Mapping the Semantic Spaces of Aesthetic Concepts. In: *International Journal of Multidisciplinary Thought*. 07(02), 295–306.

Démuth, A. (2019). *Beauty, Aesthetic Experience, and Emotional Affective States*. Berlin: Peter Lang Verlag.

Démuth, A. (2020). K problematike geometrizovania krásy a estetických pojmov. [On the issue of geometrizing beauty and aesthetic concepts.] *Filozofia*. 75, 2, 121–132. https://doi.org/10.31577/filozofia.2020.75.2.4

Derksen, R. (2008). *Etymological Dictionary of the Slavic Inherited* Lexicon (Leiden Indo-European Etymological Dictionary Series; 4), Leiden, Boston: Brill.

de Vaan, M. (2008). *Etymological Dictionary of Latin and the other Italic Languages*. (Leiden Indo-European Etymological Dictionary Series; 4) Leiden, Boston: Brill.

DWDS – *Digitales Wörterbuch der deutschen Sprache*. Das Wortauskunftssystem zur deutschen Sprache in Geschichte und Gegenwart, hrsg. v. d. Berlin-Brandenburgischen Akademie der Wissenschaften, <https://www.dwds.de/>,

Eco, U. (2005). *Dějiny krásy*. [History of Beauty.] Praha: Argo.

Harper, D. (2020). "Beauty". In: *Online Etymology Dictionary*. dostupné na https://www.etymonline.com/search?q=beauty

Hilský, M. (1997). Shakespearove Sonety. [Shakespeare's sonnets.] In: Shakespeare, W. (ed.). *Sonety – The Sonnets*. Praha: Torst.

Hornby, A. S. (ed.) (2011). *Oxford Advanced Learner's Dictionary of Current English*. Oxford: Oxford University Press.

Hosoya, G., Schindler, I., Beermann, U., Wagner, V., Menninghaus, W., Eid, M., Scherer, K., R. (2017). Mapping the Conceptual Domain of Aesthetic Emotion Terms: A Pile-Sort Study. *Psychology of Aesthetics, Creativity, and the Arts*. 11(4), 457–473. https://psycnet.apa.org/doi/10.1037/aca0000123.

Chen, W., Qi, J., & Hao, P. (2018). On Chinese Aesthetics: Interpretative Encounter between Taoism and Confucianism. *Culture and Dialogue*. 6(1), 61–76. https://doi.org/10.1163/24683949-12340042.

Keçeli, Y., Démuth, A. & Démuthová, S. (2021). Güzellik Hakkında Konuşurken Aynı Şeyi Düşünüyor Muyuz? *Türk Dili Araştırmaları Yıllığı – Belleten*. 72 (Aralık), 301–320. Retrieved from https://dergipark.org.tr/en/pub/belleten/issue/66046/1032105.

Kluge, F. (1891). *Etymologisches Wörterbuch der deutschen Sprache*. Berlin: Walter de Gruyter.

Králik, Ľ. (2015). *Stručný etymologický slovník slovenčiny*. [A brief etymological dictionary of Slovak.] Bratislava: Veda.

Levinson, J. (2011). Beauty is Not One: The Irreducible Variety of Visual Beauty. In: Schellekens, E, Goldie P. (eds.). *The Aesthetic Mind*. Oxford: Oxford University Press, 190–207.

Liddell, H., G., Scott, R. (1940). *A Greek-English Lexicon*; Machine readable text. Oxford: Trustees of Tufts University.

McIntosh, C. (ed.) (2013). *Cambridge Advanced Learner's Dictionary*; Cambridge: Cambridge University Press.

Mennighaus, W., Wagner, V. Kegel, V. Knoop, C., A., Schlotu, W. (2019). Beauty, Elegance, Grace, and Sexiness Compared. *PLoS ONE*. 14 (6), e0218728. https://doi.org/10.1371/journal.pone.0218728

Nisanyan, S. (2011). *Sözlerin Soyağacı Çağdaş Türkçenin Kökenbilim Sözlüğü*. Istambul: Everest Yayınları.

Parkes, G., Loughnane, A. (2018). "Japanese Aesthetics", *The Stanford Encyclopedia of Philosophy* (Winter 2018 Edition), Edward N. Zalta (ed.), URL = <https://plato.stanford.edu/archives/win2018/entries/japanese-aesthetics/>.

Pfeifer, W. et al. (1993). *Etymologisches Wörterbuch des Deutschen*. Digitalisierte und von Wolfgang Pfeifer überarbeitete Version im Digitalen Wörterbuch der deutschen Sprache, dostupné na: <https://www.dwds.de/wb/wb-etymwb>

Pokorny, J. (1959). *Indogermanisches etymologisches Wörterbuch*. Bern: Francke.

Rejzek, J. (2001). *Český etymologický slovník*. [Czech etymological dictionary]. Voznice : Leda.s. a. (2015). *Gramatika súčasnej gréčtiny*. [Grammar of modern Greek] Bratislava: Lingea.

Sartwell, C. (2004) *Six Names of Beauty*. New York: Routledge.

Schelling, F. W. J. (1800/1907): System des transcendentalen Idealismus, *Werke*. Band 1, Leipzig: Fritz Eckart.

Schiller, F. (2005): On Grace and Dignity. In: Curran, J. V., Fricker, Ch. (eds.): *Schiller's "On Grace and Dignity" in Its Cultural Context: Essays and a New Translation. Studies in German literature, Linguistics, and Culture*. Rochester, NY: Camden House, 152 – 153.

Špaňár, J., Hrabovský, J. (1998). *Latinsko/slovenský-Slovensko/latinský slovník*. [Latin/Slovak-Slovak/Latin dictionary]. Bratislava: Slovenské pedagogické nakladateľstvo.

Vilchez, J., M., P., (2017). *Aesthetics in Arabic Thought: From Pre-Islamic Arabia Through Al-Andalus*. Handbook of Oriental Studies. Section One: The Near and Middle East. Leiden: Brill.

Vopěnka, P. (2001). *Úhelný kámen evropské vzdelanosti a moci*. [The cornerstone of European education and power]. Praha: Práh 2001.

Walde, A. (1938). *Lateinisches Etymologisches Woerterbuch*. 3e Auflage, bearb. bei J. B. Hoffmann. Heidelberg: C. Winters Universitätsverlag.

Żmigrodzki, P. et al. (2018). *Wielki słownik języka polskiego*. [A large dictionary of the Polish language]. Warszawa: Instytut Języka Polskiego PAN. ttps://www.wsjp.pl/index.php?szukaj=ładny.

The Possibilities of Studying Connotations of the Term "Beauty" in a Natural Language

Slávka Démuthová

Abstract. The use and subsequent understanding of the language used for communication necessitates a certain consensus on what is meant by individual concepts. Learning a language is to become familiar with the meanings of individual words, learning what a word represents in the given culture, society, community or group. A large part of the vocabulary of a language comes from the ability to use words with the meaning that is commonly understood by the majority of people we wish to communicate with during our lives. However, there is still a rather large group of (mostly abstract) words which do not have an entirely clear meaning, or that have significant differences in meaning from one individual to another. Such broadly based terms used in a language include the term "beauty". Its content is highly individual, determined by personal experience, aesthetic preferences, cultural influences, contemporary context, and age and sex. Notwithstanding this fact, this term, which is frequently employed, is not only used in general communication but is also a key term in numerous scientific fields, such as art, philosophy, aesthetics, psychology and medicine. To work with this key term we need to continuously study it and search for its meaning. One effective method used to study the understanding of the term "beauty" is the study of its connotations in a natural language (ordinary language). The following text presents several of the possibilities that are currently available for such studies and provides the outcomes from recently published studies that have dealt with the identification of the connotations of the term "beauty" in a natural language.

Introduction

Beauty is one of the most broadly based terms used in a language (Kenett et al., 2021). It is related to aesthetic perception and a subsequent evaluation, which are subject to contemporary, social, cultural... but especially individual preferences (Jacobsen, 2010; Démuth, 2019). Understanding this term is difficult, not only due to its variability but also due to its multidimensional character. The mechanisms and components of evaluation that are used when we declare that something or somebody is beautiful differ to such an extent that it is hardly possible to define or accurately describe the term "beauty". Even the previously unambiguous attribute, the presence of positive emotions (Juslin, 2013), is no longer certain – the multidimensional character of this concept can be seen when beauty is attributed to situations when one experiences fear or sorrow (Ishizu & Zeki, 2017) or if something is so ugly that it is considered to be beautiful. Beauty is a concept and variable that is subject to examination in a number of different disciplines: not only in art (Sidhu et al., 2018), psychology (Yarosh, 2019) and medicine (Corbett, 2009), but also in mathematics (Zeki et al., 2018), philosophy (Scruton, 2011) and biology (Jones & Jaeger, 2019). But none of these disciplines has given up their investigation because of the complexity or variability of the concept of beauty. On the contrary, these characteristics, inter alia, allow the application of different approaches and viewpoints to the study of beauty (scientific or artistic) – each individual method is able to clarify a certain aspect of the complex term "beauty" better than another.

One of the approaches that reflects the current situation (contemporary context), the cultural and social specificities (regional context) as well as individual preferences related to beauty is the examination of the use of a selected term in a natural language. It can provide information on who uses the studied term, when and how often they use it, how they combine it with other terms, their collocations, etc. These quite complicated analytical processes used to study the occurrence of a given term in a natural language require the study of many of the levels within a language (lexical, semantic, pragmatic, discursive... Egorova et al., 2013); hence, it is often impossible to do this without the help of sophisticated computer programs or artificial intelligence (see e.g., Mikolov et al., 2013; Peng et al., 2020; Reading Turchioe et al., 2022). Another, more accessible, method of study into the use of the term "beauty" in a natural language is an analysis of its connotations (see e.g., Courts & Bartol, 1996; Manchaiah et al., 2015; Page et al., 2021). Here, a study of the connotations of the term "beauty" in a natural language (whether in its oral form or (more frequently) in written texts) is based on the so-called "distributional hypothesis" (Harris, 1954), which assumes that expressions that connote a specific term tend to occur in similar linguistic contexts (Kastrin & Hristovski, 2019). A study of the connotations of a term is not only used with a view of the mutual proximity of connotations in texts (or oral utterances) but mainly because the connotations of a term reflect one's experience of the specific term; hence, the content of the term "beauty" is often expressed very accurately. This method is also based on the assumption that an internal definition of a term (i.e., how one understands (defines it for themselves) the given term) is observable in what further expressions the person associates with the stimulus word. Not only in respect to other objects but also emotions, attributes, ideas or even (implicit) experiences that are expressed with difficulty.

Thus, a survey of the meaning of the term "beauty" can be carried out using a number of methods; some of them are quite complicated (i.e., computer-based techniques working in virtual spaces (e.g., a computational semantic approach – Mikolov et al., 2013) or theoretical multidimensional models, which, in many respects, are only waiting for their practical use (semantics based on conceptual spaces – Gardenfors, 2014)), while others can, with little difficulty, be directly applied in field research (e.g., a frequency analysis of connotations – Démuthová & Démuth, 2021). Their common denominator is that they bring a full meaning of the term "beauty", dependent upon the data obtained from its use in the specific natural language. After some simplification, they can be divided into two categories, depending on the number of domains they track, single (e.g., frequency analysis) and multidimensional (e.g., semantic differential).

Single-Dimensional Methods

Single-dimensional methods only track a single domain of connotations of the term "beauty". They can be focused on the frequency of occurrence of a given connotation in the group/population studied, or, for example, on an analysis of content or morphological and other characteristics of the connotations that are produced (or the approaches can be combined). They cannot determine the position of the connotations (and ultimately the position of the term "beauty" itself) in the semantic space (whether it be two, three or multi-dimensional); however, they still have the potential to convey valuable information, especially where the basic research in this field has not yet been carried out.

Frequency Analysis of Connotations

Frequency analysis monitors the multiplicity (frequency) with which the given connotation occurs in the population studied, especially in comparison to other connotations. This method allows the expressions, thoughts, ideas and emotions... that are most frequently associated with the term "beauty", to be identified within a study sample. At first sight it is a very simplistic method; yet it not only has the potential to find the most frequent connotations of any selected term but also, for example, to evaluate which are the most typical connotations reported by a group of participants based on the order in which the connotations occurred. It can be assumed that those connotations that were identified first are those that are closest to the term "beauty", capturing it better than the connotations that came up at the end of the list.

An example of this type of research was a study that used a sample of over 1,500 Slovak participants between the ages of 19 to 89 (average age = 45.23, SD = 16.33 years), 52.7 % of the participants were female (Démuthová & Démuth, 2021). The most frequent connotations of the term "beauty" included nature (reported by 35.28 % of participants), woman (reported by 21.73 % of participants), love (17.77 %), family and children (both 15.02 %). On the contrary, the end of the list (only connotations with a frequency that exceeded 1 % were included in the analysis, as associations below the level of 1 % are considered to be highly original, and thus atypical – Said-Metwaly et al., 2021) included feeling, freedom, water (1.27 %), landscape and Xmas (both 1.21 %), figure (1.07 %) and heart (1.01 %) (Démuthová & Démuth, 2021). Subsequent analysis of the content (see the following sub-chapter) can be used to interpret the meaning of the most frequent connotations, and thus the meaning of the term "beauty".

However, further studies are possible within the frequency analysis, such as a survey of the occurrence of connotations with respect to various characteristics (age, sex, education, etc.). It can be assumed that the content of the term "beauty" can change with age. Research has also shown that there are differences between the sexes. For instance, women (unlike men) ranked the word "grandchild" quite high on the list of connotations; on the contrary, men (unlike women) more often included the terms "animal", "friend", "forest" and "peace" (Vavrová & Démuthová, 2021). These differences can be interpreted in the context of a greater degree of engagement of women in family life and with their children and grandchildren (Stelle et al., 2010). With men it can be seen in the context of spending greater amounts of leisure time in comparison to women (Azevedo et al., 2007).

The frequency analysis of connotations of the term "beauty" can also be used to monitor the occurrence of the parts of speech. Morphological characteristics is additional information that allows a better and more accurate understanding of the term "beauty". It can be assumed that the dominance of nouns among the reported connotations would point to the possibility, in a natural language, that beauty is substantiated and based on an object. The dominance of adjectives might indicate that the essence of beauty lies in the quality (grade) of the subjects, ideas and objects. The extensive occurrence of verbs among the connotations might suggest that beauty has the potential to energise individuals and spur them to action. The need for research in this field is supported by the outcomes of our analyses (see Démuthová & Démuth, 2021), showing that up to 93.7 % of connotations were nouns. This dominance of nouns is surprisingly high. It appears that when connotations are produced and after a search for the words closest to the studied concept, there is a significant tendency to maintain a grammatical closeness to the stimulus word; thus, if it is a noun, the participants will primarily produce connotations that are nouns. This hypothesis could be verified through an analogous piece of research. If the stimulus word was, for example, "beautiful" and adjectives were dominant in the reported connotations (or if their occurrence was significantly higher than had been reported for the stimulus word "beauty"), then it would be clear that the morphological overlay (burden) of the stimulus word (notion) is dominant in the connotations reported and information on the frequency of the separate parts of speech among the connotations has no informative value as to whether the term "beauty" is substantiated, or if it merely reflects the quality (grade) of the subjects, phenomena, ideas, etc. under consideration.

Content Analysis of Connotations

Content analysis is focused on the meaning of connotations related to the term "beauty". In a similar way to the meaning analysis of the inter-gender differences in reported connotations of the term "beauty" (the orientation of women towards family and men towards leisure time), an analysis of the general contents of the term "beauty" through its connotations can also be attempted. Content analysis aims to provide information on what is most frequently "on one's mind" in association with the term "beauty". Using other words, it depicts the objects, states or mental representations that are associated with the term "beauty". Content analysis is typically the second step in the hierarchy it lies above frequency analysis, as it conveys those contents that are most frequently associated with the term "beauty". Unlike frequency analysis, it is accompanied by a certain degree of generalisation, as it requires a meta-analysis of the information in order to identify the common features, focus and areas that are associated with the most common connotations (or group of connotations) to enable them to be characterised as a group. Contrary to the reasonably objective data obtained from a frequency analyses, the methods used in a content analysis generate a certain degree of subjectivity through the necessity to determine one's own interpretational framework. Despite these limits the content analysis of connotations has an irreplaceable place in the methods that aim to clarify the meaning of this term in a natural language.

A content analysis of 63 connotations was carried out in our research. They included expressions with a frequency of occurrence greater than 1 % for the study sample (from frequency analysis), using the above-mentioned criterion, i.e. any associations below this limit are considered to be highly original, and thus atypical (Said-Metwaly et al., 2021). The 63 connotations included very specific terms (car, book, eyes...) as well as significantly abstract notions (heaven, soul, purity...). In an effort to put these terms into suitable categories, seven categories (see Table 1) were identified and given umbrella terms.

At this level it can be stated that the term "beauty: is most frequently associated with notions that denote nature, values, feelings,

Table 1 Beauty connotations categorised based on their content

Categories	Connotations					
Nature (n=12; 19.05 %)	nature*	flower*	sun*	sea	forest	snow
	water	sky	landscape	mountain	tree	animal*
Values (n=12; 19.05 %)	love*	family*	life*	home	naturalness	purity
	peace	freedom	health	perfection	good (n)	youth
Feelings (n=9; 14.29 %)	joy	pleasant	happiness	feeling	inner	
	natural	relax	gorgeousness	nice		
People (n=9; 14.29 %)	me	woman*	friend	child*	human*	
	mother	grandchild	man	wife		
Objects (n=9; 14.29 %)	car*	book	picture	clothes	face	
	figure	eyes	fashion	smile		
Abstract Ideas (n=8; 12.70 %)	God	heaven	soul	Xmas		
	art*	aesthetics	heart	music*		
Activities (n=4; 6.35 %)	sex	sport	work	vacation		

Notes: (n) = nouns; *= a connotation with a frequency greater than 5 %
Source: Démuthová & Démuth (2021, 7)

people, objects and activities, as well as abstract ideas in a natural language. Although the content analysis has the potential to clarify the meaning of the term "beauty", a number of essential characteristics still remain hidden. The basis of the interconnections that are responsible for the association of the term "beauty" and its connotations remain hidden, and there is also no distinction made in terms of the intensity of the connections, etc. These characteristics can be revealed through other methods, some of which are described in the following paragraphs.

Multidimensional Methods

Multidimensional methods give the possibility to monitor several features (characteristics) of the given term at once. In this way they provide a more comprehensive and more accurate view of the studied term. Methods that apply several characteristics/dimensions of a term can be quite accurately compared to data on the location of a particular place on Earth – a single dimension (e.g., longitude) is a rather broad definition, while two dimensions (e.g., longitude and latitude) provide a more accurate location, but only in two-dimensional space. Three dimensions would allow the addition of further data, for example, the location's altitude. The considerations related to such a coding system do not have to be limited by the dimensions of the space; they can be extended by others (such as time), which are essential to the depiction and characteristics of a place (in this case, the location data is complemented by information on contemporary context). We can say that the more meaningful dimensions we are able to use, the more accurate the description of the place. In addition, the application of dimensions to other places allows us to obtain valuable information on the location of additional places, which can subsequently be compared. In this way, we can study the mutual relationships between places, e.g., it can be determined whether another place is located in the close vicinity or if it is very far away, or in which dimensions it is similar and in which dimensions it differs.

Analogously, the position of any word, expression or notion in the so-called "semantic space" can be considered. In this case the greatest difficulty is in finding suitable dimensions which, if applied, would allow relevant information to be obtained on the given notion and would also define the semantic space. It would even be optimal to identify the dimensions that are applicable, not only to the given term but also to various additional terms to allow the positioning of other expressions in the semantic space (and possibly to compare their closeness within the individual dimensions). Such an analogous use of spatial proximity, for more abstract entities such as terms (compared to physical locations, the location of a particular place), i.e. the concept of semantic spaces can be used to study the characteristic features of various terms and their related expressions, is based on the assumption that the meaning of a word is determined by its context and that words with similar meanings appear in similar contexts (de Boer et al., 2018). This proximity was already referred to by J. R. Firth, in 1957, in his well-known quote: "You shall know a word by the company it keeps." Semantic space basically refers to the way that terms are organised in our internal vocabulary, in our internal dictionary that we use to save and search for words and their meanings. This organisation of words is often conceptualised as a space with several dimensions, in which words are represented as points or vectors and the distances between the points reflect the semantic relationships between words. For instance, semantically similar words would be represented as points that are close to each other, while semantically different words would be represented as points that are distant from each other. Context is then defined as a set of words that are closely grouped in semantic space (Boer et al., 2018). Several possibilities and methods may be applied to the understanding of terms in semantic space. One of the oldest methods, focused on the identification of a location for individual words in dimensions of the semantic space, is the semantic differential method (Osgood, 1957; Osgood, Suci, & Tannenbaum, 1957; Snider & Osgood, 1969).

Semantic Differentials

The semantic differential method allows measurements of the subjective evaluations of individuals related to a particular term through a series of bipolar adjectives. In practice, this means that an individual evaluates a selected term on a scale (typically seven-point) based on their conviction with regards to how close or far the given term is to or from one of two opposite adjectives (e.g., nice – ugly, fast – slow, etc.). On the basis of this evaluation, the term can be given a value on a particular scale (e.g., on the beauty scale, speed scale, etc.). This can subsequently be depicted in the space between two points that correspond to two counterpoints – opposite adjectives. This method has the advantage (e.g., compared to the monitoring of connotations) that it makes the information on the character of the relationship between two terms (the evaluated term and an adjective) accessible, the results are quantifiable (how close the term is to its adjective), and at the same time, due to scale bipolarity, both variants of a relationship – positive and negative – are measured. This basic level of semantic differentials allows work with a large amount of data – the results can reveal, for a selected group of individuals, the general public or even a specific culture, how they understand the evaluated term in the context of selected criterion (i.e., the nice – ugly evaluation). We can compare individuals, groups, communities or cultures according to gender (see e.g., Kenett et al., 2021), age (Menninghaus et al., 2019) or other criteria. The use of several scales (bipolar adjectives) allows more comprehensive and plastic information on the characteristics and nature of an evaluated term to be obtained. It is possible to study which adjectives provide similar results for a particular study group and which provide variable results. It is also possible to consider which adjectives are suitable to be used to describe a selected term (frequently associated with the given term), and which, on the contrary, provide such variation that their connection to the evaluated term is highly individual, hence atypical. Extreme (either very high or very low) values point to a very close connection, the inclination of the evaluated term to the given adjective, thus highlighting the essential characteristics associated with that term.

An evaluation of the term "beauty" through adjectives was also conducted in a study of the Slovak population. A total of 48 bipolar adjectives were used, and for several of them values close to the limit values were reported. Table 2 shows the bipolar adjectives with the average values obtained from a group of 1,563 participants (52.7 % were female) aged 12–98 (average age = 45.2), which were the most polarised.

It is clear from the data presented in Table 2 that the majority of participants imagine beauty as something that is highly pleasant, beautiful, good and, at the same time, inspiring, tempting and attractive. Similar results were reported by W. Menninghaus et al. (2019), who identified boundary values in relation to the term "beauty" and the adjectives tasteful, pleasant, and natural.

It is not only in this case, but especially in relation to terms that are characterised by several different adjectives, that we encounter difficulties in characterising such adjectives at a general level. Hence, again, a need arises to categorise the characteristic adjectives, to find factors of higher degree, dimension that might cover them, that represents a problem similar to that which appeared in

Table 2 Bipolar adjectives used to describe the term "beauty" that reported average values at the boundaries (<2.5 and >5.5)

	1	2	3	4	5	6	7	
Pleasant	1.83							Unpleasant
Good		2.21						Bad
Beautiful	1.96							Ugly
Natural		2.40						Artificial
Repugnant						5.68		Attractive
Clean		2.43						Dirty
Inspiring		2.36						Non-inspiring
Tempting		2.27						Repulsive

Source: author

the search for meta-categories in the content analysis of the frequent connotations of a term. Therefore, another advantage of the use of semantic differentials is that it does not end in an analysis of term evaluated through a scale of bipolar adjectives. The method of semantic differentials includes various advantages such as the use of a standardised method for the measurement of the meanings of words and notions. It provides the possibility to use a bipolar scale to evaluate both the positive and negative connotations of terms and a numeric scale that allows an accurate measurement of the response given by participants. The evaluation scale thus measures the connotative meaning of terms and, unlike other evaluation scales, it is universal, i.e., it can measure associations (Xiong et al., 2006), attitudes (Heise, 2007), motivations and emotions (Bradley & Lang, 1994) ..., for almost any concept (Ploder & Eder, 2015). This type of data enables further meta-analysis using factor analysis. The author of semantic differentials, Charles E. Osgood, identified three systematically repetitive and relatively stable dimensions using a factor analysis for the answers (evaluations) provided for different words on the bipolar scales of adjectives. He called them: evaluation (focused on the value of the object), potency (referring to the power of an object) and activity (describing the movement of an object), represented by the series of adjectives (see Table 3) - e.g., good/bad (evaluation), strong/

weak (potency) or slow/fast (activity) (Rosenberg & Navarro, 2018). These three dimensions were "found to underlie most of the semantic space" (Osgood, 1957, 390).

Table 3 Examples of commonly used adjective pairs for each dimension of semantic differentials

Evaluation	Potency	Activity
Good/bad	Strong/weak	Passive/active
Expensive/cheap	Decisive/indecisive	Lazy/industrious
Wise/foolish	Hard/soft	Aimless/motivated
Beautiful/ugly	Potent/impotent	Calm/excitable
Honest/dishonest	Severe/lenient	Fast/slow
Kind/unkind	Brave/cowardly	Emotional/unemotional

Source: Rosenberg & Navarro (2018, 1504)

Factor analysis thus enables an analysis of the proximity or distance between the ratings of the adjectives and thus determines the different dimensions that represent them in the semantic space (Ploder & Eder, 2015). Three dimensions (evaluation, potency and activity) can be used as three vectors, thus defining a space in which an evaluated concept (or several concepts) can be positioned, along with a graphical presentation of their position in semantic space (see Picture 1).

Several studies consider the dimensions of evaluation, potency and activity as sufficiently broad. They cover 50–70 % of the connotations when semantic differential is applied (Skrandies, 2011); however, there are studies that underline the need to adapt the

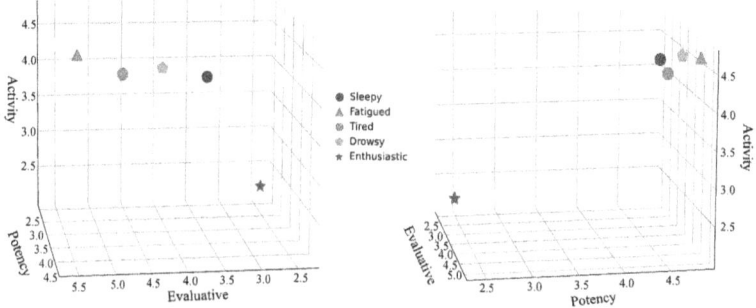

Picture 1 An example of the depiction of similar terms in semantic space.
Source: Long et al. (2022, 1239)

dimensions to a specific term. At the same time, they appreciate a certain degree of openness that provides the opportunity to use other dimensions to depict the terms under study (see e.g. Démuth, 2017). As early as the seventies, the same method (factor analysis) was used by Mehrabian and Russell (1974) and again three dimensions were discovered. This time they were pleasure, arousal and dominance. Quite extensive use of these three domains to define the semantic space of various terms led to extensive research that resulted, inter alia, in the creation of the ANEW (Affective Norms for English Words) lexicon. The lexicon is made up 2,471 English words with their average values and statistical deviations, as the results of the assessment of the individual words by individuals on three scales: pleasure, arousal and dominance (Bradley & Lang 1999). The greatest degree of saturation of the domain pleasure was found for the terms "beauty" and "beautiful" (see Table 4).

Table 4 Mean and standard deviation for the domains of pleasure, arousal and dominance for the words "beauty" and "beautiful" as rated in the ANEW (averaged across all, male, and female subjects)

	Dimensions	Pleasure		Arousal		Dominance	
Word	Subjects	Mean	SD	Mean	SD	Mean	SD
Beauty	All	7.82	1.16	4.94	2.57	5.53	2.10
	Male	7.47	1.13	5.6	2.16	5.33	1.50
	Female	8.04	1.15	4.52	2.76	5.65	2.44
Beautiful	All	7.60	1.64	6.17	2.34	6.29	1.81
	Male	7.23	2.00	6.36	2.46	5.65	2.44
	Female	8.00	1.03	5.95	2.24	6.50	2.01

Source: Bradley & Lang 1999, 19, 33

Not all procedures have led to the identification of three dimensions, for example, through a factor analysis of the results from a semantic differential, using 42 adjectives that describe terms related to the topic of beauty (beauty, elegance, grace, and sexiness), the study we have already mentioned by W. Menninghaus et al. came up with seven domains that covered their evaluation: tastefulness/fluency, affective habitus/arousal, cultural sophistication, intellectual rigor, lightness/slenderness, exquisiteness and delicacy/femininity (Menninghaus et al., 2019). The larger the number of domains the better the common "meta-characteristics"/aspects of the features that are depicted by the adjectives, but on the other hand it is very complicated to integrate them into the concept of semantic space. The simultaneous assessment of several dimensions, which is sufficiently complicated, can be simplified with the use of computer techniques. The most frequently used methods include the computational semantic approach and latent semantic analysis.

Latent Semantic Analysis

Computer-based techniques are a variant of the described analyses of the use of a natural language and the subsequent modelling of semantic spaces for more complex forms. An example of such an approach is latent semantic analysis (LSA). LSA is a method for the extraction of meaning from text based on statistical computations applied to a collection of documents (Evangelopoulos, 2013) such as a sentence, paragraph or essay (Landauer et al., 1998). The first step in LSA is the creation of a matrix that contains the frequency of each word/term in the documents under study. In such a matrix, the rows are the words in the corpora, the columns represent the documents and the cells include the number of occurrences of a word (frequency). Such a matrix represents the so-called "LSA Space" (Vrana et al., 2018), where documents are represented as vectors in the "term space" and at the same time, terms are represented as vectors in the "document space". The extent to which documents and words correspond is revealed through the use of cosine similarity, defined as the cosine of the angle formed by these two vectors (Evangelopoulos, 2013). This matrix is then transformed, through "singular value decomposition", a method of data reduction often referred to as principal component analysis. It is a mathematical approach that decomposes a matrix into three smaller matrices: a text vs dimension matrix (giving the positions of the texts in the semantic space); a word vs dimension matrix (used to find the position of additional texts in the semantic space); and a matrix of the "singular values" of each dimension (showing what fraction of the total variance is captured by that dimension) (Vrana et al., 2018). The output is a highly dimensioned space that can comprise of tens up to hundreds of dimensions (Landauer et al., 1998). The processing of such data is made possible by computer programs. The main advantage of LSA is the ability to transform language into a quantitative vector. In this context, LSA is not only an analytic method but also a theory that explains how humans acquire, induct, and represent meanings and knowledge (Landauer & Dumais, 1997). Therefore, the outcomes of the analyses can be used in different tasks relating to processing of natural languages, such as text classification, the search for information and data visualisation in the semantic space. With regard to the term "beauty", LSA can be used not only to depict the meanings of this term in various corpora but also to analyse the similarities between the term "beauty" and other (similar) terms, or to analyse the similarities or differences between documents.

Computational Semantic Approach

The computational semantic approach is a method which, according to several authors, outperforms traditional computational semantic models, such as latent semantic analysis (Mikolov et al., 2013). A disadvantage of latent semantic analysis is that it makes no use of word order and is thus incapable of identifying syntactic relationships (de Boer et al., 2018). The computational semantic approach is able to understand the meaning of a text in a natural language through the use of computer algorithms and mathematical models. It can perform an analysis of the syntactic structure of sentences and use this information to derive their basic semantic meaning. Another advantage is that such approaches use machine-learning models, which are able to not only understand a language but also generate it. This is done by teaching the models using large amounts of annotated text data, which allows them to learn the patterns of the relationships between words and phrases.

An example of the application of computer-based techniques in the field of beauty is the study by Kenett et al. (2021), who focused on the similarities and differences between the similar terms "beauty" and "wellness" in different age groups and gender. In the first phase they used data from the Word2vec program, which is focused on word-level textual corpora, trained on a corpus of 100 billion words taken from Google News over 10 years (Rezaeinia et al. 2017). Word2vec understands and assigns vectors to the meaning of words in any document. It does this based on the presumption, common for the majority of semantic space methods, that words with similar meanings in a given context are close to each other (Sahlgren, 2008). Using this database, the authors used the first 10 % (300,000) of words with the highest frequency of occurrence and subsequently derived approximately 700 word vectors, which were those that were most similar (cosine similarity) to the word vector Beauty. Subsequently, they chose the expressions that were typical of this term but had not occurred in relation to other similar words. Hence the term "beauty" has been shown to be consistently related to words such as elegance, feminine, gorgeous, lovely, sexy and stylish (Kenett et al. 2021). The authors also studied how the understanding of the term "beauty" changes depending on the age and gender of a participant, as it is clear that these two variables can significantly affect the understanding of this term (see e.g., Vavrová & Démuthová, 2021; Wulff et al., 2022). Their findings included the fact that within the semantic networks of the older cohorts the concepts of beauty and wellness were less connected to each other and were more widely segregated. Also, the degree of segregation and organisation of concepts was greater in women in comparison to men (Kenett et al. 2021).

The use of the computational semantic approach does not end once the analyses of terms and texts are complete; it has a great potential, it can be used in a broad range of tasks relating to the processing of natural languages, such as text classification, information extraction, machine translation and answering questions. It is also able to generate text in a natural language, which can be used, for example, in the creation of chatbots, virtual assistants and other interactive systems, which can understand and react to inputs in natural languages (Ahram, 2019).

Conclusion

Understanding the term "beauty" and its subsequent definition is complicated by several factors. It is one of the most broadly based terms used in a language. It is closely related to aesthetic perception and a subsequent evaluation, which is affected by numerous factors. To identify the characteristics of this term thus necessitates the consideration of a number of variables, which makes the creation of a complex notion complicated and unclear. Help with the visualisation of various relationships can lead to a transformation of the mental space, where terms are shaped into the so-called "semantic space". This transformation is based on the concept that it is possible to represent the meaning of single words as continuous mathematical objects. Semantic space models then attempt to represent the word meanings by representing them as points in an abstract (multi)dimensional space. There are relatively simple methods, which study one or only a few dimensions, but also significantly complicated modelling methods, working with anything up to many hundreds of dimensions. Advancements in computer modelling have removed many obstacles and created almost unlimited possibilities. Not only with respect to this rapidly developing potential but mainly due to the possibility that complex terms may be explored through other words (connotations) that they are associated with, through various relationships. Thus computer modelling provides a rich source of opportunities for research into mapping of the term "beauty" in natural languages.

References

Ahram, T. (ed.). (2019). Advances in artificial intelligence, software and systems engineering. *Proceedings of the AHFE 2019 International Conference on Human Factors in Artificial Intelligence and Social Computing, the AHFE International Conference on Human Factors, Software, Service and Systems Engineering, and the AHFE International Conference of Human Factors in Energy*, July 24–28, 2019, Washington D.C.: Springer International Publishing. https://doi.org/10.1007/978-3-030-20454-9

Azevedo, M., R., Araújo, C., L., Reichert, F., F., Siqueira, F., V., da Silva, M., C., & Hallal, P. C. (2007). Gender differences in leisure-time physical activity. *International journal of public health*, 52(1), 8–15. https://doi.org/10.1007/s00038-006-5062-1

Bradley, M., M., & Lang, P., J. (1994). Measuring emotion: the self-assessment manikin and the semantic differential. *Journal of Behavior Therapy and Experimental Psychiatry*, 25(1), 49–59. https://doi.org/10.1016/0005-7916(94)90063-9

Bradley, M., M., & Lang, P., J. (1999). Affective norms for English words (ANEW): Instruction manual and affective ratings. *Technical Report C-1*, 30(1), 25–36.

Corbett, J., R. (2009). What is beauty?: Royal Victoria Hospital, Wednesday 1st October 2008. *The Ulster Medical Journal*, 78(2), 84–89. https://doi.org/https://www.ncbi.nlm.nih.gov/pmc/articles/PMC2699193/

Courts, N., F., & Bartol, G., M. (1996). Psychosomatic: Connotations for people who are neither nurses nor physicians. *Clinical Nursing Research*, 5(3), 283–293. https://doi.org/10.1177/105477389600500304

de Boer, J., N., Voppel, A., E., Begemann, M., J., H., Schnack, H., G., Wijnen, F., & Sommer, I., E., C. (2018). Clinical use of semantic space models in psychiatry and neurology: A systematic review and meta-analysis. *Neuroscience and Biobehavioral Reviews*, 93, 85–92. https://doi.org/10.1016/j.neubiorev.2018.06.008

Démuth, A. (2017). Conceptual analysis of the concept of beauty in cognitive-scientific research. In Démuth, A. (ed.). *The Cognitive Aspects of Aesthetic Experience – Introduction* (31–52). Frankfurt am Main: Peter Lang.

Démuth, A. (2019). *Beauty, Aesthetic Experience and Emotional Affective States*. Berlin: Peter Lang.

Démuthová, S., & Démuth, A. (2021). A frequency and semantic analysis of the most frequent connotations of the notion of beauty. *European Journal of Behavioral Sciences*, 4(1), 1–11. https://doi.org/10.33422/ejbs.v4i1.611

Egorova, N., Shtyrov, Y., & Pulvermüller, F. (2013). Early and parallel processing of pragmatic and semantic information in speech acts: Neurophysiological evidence. *Frontiers in Human Neuroscience*, 7, 86. https://doi.org/10.3389/fnhum.2013.00086

Evangelopoulos N., E. (2013). Latent semantic analysis. Wiley interdisciplinary reviews. *Cognitive Science*, 4(6), 683–692. https://doi.org/10.1002/wcs.1254

Firth, J., R. (1957). *A Synopsis of Linguistic Theory. Studies in Linguistic Analysis*. Oxford: Blackwell.

Gardenfors, P. (2014). *The Geometry of Meaning: Semantics Based on Conceptual Spaces*. Cambridge, London: MIT Press. https://doi.org/10.7551/mitpress/9629.001.0001

Harris, Z., S. (1954). Distributional structure. *Word*, 10, 146–162. https://doi.org/10.1080/00437956.1954.11659520

Heise, D., R. (2007). *Expressive Order. Confirming Sentiments in Social Actions*. Heidelberg: Springer-Verlag.

Ishizu, T., & Zeki, S. (2017). The experience of beauty derived from sorrow. *Human Brain Mapping*, 38(8), 4185-4200. https://doi.org/10.1002/hbm.23657

Jacobsen, T. (2010). Beauty and the brain: Culture, history and individual differences in aesthetic appreciation. *Journal of Anatomy*, 216(2), 184-191. https://doi.org/10.1111/j.1469-7580.2009.01164.x

Jones, A., L., & Jaeger, B. (2019). Biological Bases of Beauty Revisited: The effect of symmetry, averageness, and sexual dimorphism on female facial attractiveness. *Symmetry*, 11(2), 279. https://doi.org/10.3390/sym11020279

Juslin, P., N. (2013). From everyday emotions to aesthetic emotions: towards a unified theory of musical emotions. *Physics of life reviews*, 10(3), 235-266. https://doi.org/10.1016/j.plrev.2013.05.008

Kastrin, A., & Hristovski, D. (2019). Disentangling the evolution of MEDLINE bibliographic database: A complex network perspective. *Journal of Biomedical Informatics*, 89, 101-113. https://doi.org/10.1016/j.jbi.2018.11.014

Kenett, Y., N., Ungar, L., & Chatterjee, A. (2021). Beauty and wellness in the semantic memory of the beholder. *Frontiers in psychology*, 12, 696507. https://doi.org/10.3389/fpsyg.2021.696507

Landauer, T., K., & Dumais, S., T. (1997). A solution to Plato's problem: The latent semantic analysis theory of acquisition, induction, and representation of knowledge. *Psychological Review*, 104(2), 211-240. https://doi.org/10.1037/0033-295X.104.2.211

Landauer, T., K., Foltz, P., W., & Laham, D. (1998). An introduction to latent semantic analysis. *Discourse Processes*, 25(2-3), 259-284. https://doi.org/10.1080/01638539809545028

Long, H., Scott, H., & Lack, L. (2022). Sleepy, tired, drowsy, and fatigue have different meanings for a university student sample. *Journal of Clinical Sleep Medicine: JCSM: Official Publication of the American Academy of Sleep Medicine*, 18(5), 1235-1241. https://doi.org/10.5664/jcsm.9780

Manchaiah, V., Stein, G., Danermark, B., & Germundsson, P. (2015). Positive, neutral, and negative connotations associated with social representation of 'hearing loss' and 'hearing aids'. *Journal of Audiology & Otology*, 19(3), 132-137. https://doi.org/10.7874/jao.2015.19.3.132

Menninghaus, W., Wagner, V., Kegel, V., Knoop, C., A., & Schlotz, W. (2019). Beauty, elegance, grace, and sexiness compared. *PloS ONE*, 14(6), e0218728. https://doi.org/10.1371/journal.pone.0218728

Mikolov, T., Sutskever, I., Chen, K., Corrado, G., & Dean, J. (2013). Distributed representations of words and phrases and their compositionality. *arXiv*. https://doi.org/10.48550/arXiv.1310.4546

Osgood, C., E. (1957). Measurement of meaning. *Psychological Bulletin*, 54(4), 390-407.

Osgood, Ch., E.; Suci, G., & Tannenbaum, P. (1957). *The Measurement of Meaning*. Urbana, IL: University of Illinois Press.

Page, S., A., Collisson, B., A., Godley, J., Nguyen, D., Metz, L., & Muruve, D. (2021). How semantics connotations may influence concerns about donation of biospecimens. *Biopreservation and Biobanking*, 19(3), 156–162. https://doi.org/10.1089/bio.2020.0072

Peng, J., Zhao, M., Havrilla, J., Liu, C., Weng, C., Guthrie, W., Schultz, R., Wang, K., & Zhou, Y. (2020). Natural language processing (NLP) tools in extracting biomedical concepts from research articles: A case study on autism spectrum disorder. *BMC Medical Informatics and Decision Making*, 20(Suppl 11), 322. https://doi.org/10.1186/s12911-020-01352-2

Ploder, A. & Eder, A. (2015). Semantic differential. In Wright, J. D. (ed.), *International Encyclopedia of the Social & Behavioral Sciences*. (Second Edition) (563–571), Amsterdam: Elsevier. https://doi.org/10.1016/B978-0-08-097086-8.03231-1

Reading Turchioe, M., Volodarskiy, A., Pathak, J., Wright, D. N., Tcheng, J. E., & Slotwiner, D. (2022). Systematic review of current natural language processing methods and applications in cardiology. *Heart (British Cardiac Society)*, 108(12), 909–916. https://doi.org/10.1136/heartjnl-2021-319769

Rezaeinia, S., M., Ghodsi, A., & Rahmani, R. (2017). Improving the accuracy of pre-trained word embeddings for sentiment analysis. *arXiv*:1711.08609. https://doi.org/10.48550/arXiv.1711.08609

Rosenberg, B., D., & Navarro, M. A. (2018). Semantic differential scaling. In B. B. Frey (ed). *The SAGE Encyclopedia of Educational Research, Measurement, and Evaluation* (1504–1507). Thousand Oaks, CA: SAGE Publications. http://dx.doi.org/10.4135/9781506326139.n624

Sahlgren, M. (2008). The distributional hypothesis. *The Italian Journal of Linguistics*, 20(1), 33–53.

Scruton, R. (2011). *Beauty: A Very Short Introduction*. (Illustrated ed.) Oxford: Oxford University Press.

Sidhu, D., M., McDougall, K., H., Jalava, S., T., & Bodner, G., E. (2018). Prediction of beauty and liking ratings for abstract and representational paintings using subjective and objective measures. *PLoS ONE*, 13(7): e0200431. https://doi.org/10.1371/journal.pone.0200431

Skrandies, W. (2011). The structure of semantic meaning: A developmental study. *Japanese Psychological Research*, 53(1), 65–76. https://doi.org/10.1111/j.1468-5884.2010.00456.x

Snider, J., G., & Osgood, Ch., E. (Eds.) (1969. *Semantic Differential Technique, A Sourcebook*. Chicago: Aldine.

Stelle, C., Fruhauf, C., A., Orel, N., & Landry-Meyer, L. (2010). Grandparenting in the 21st century: issues of diversity in grandparent-grandchild relationships. *Journal of Gerontological Social Work*, 53(8), 682–701. https://doi.org/10.1080/01634372.2010.516804

Vavrová, M., & Démuthová, S. (2021). Interpohlavné rozdiely v konotátoch pojmu krása. [Sex differences in the connotations of the concept of beauty]. In Rojková, Z., & Kochanová, D. (eds.) *Kondášove dni 2021* [Kondas' Days] (pp. 69–76). Trnava: Katedra psychológie FF UCM v Trnave.

Vrana, S., R., Vrana, D., T., Penner, L., A., Eggly, S., Slatcher, R., B., & Hagiwara, N. (2018). Latent semantic analysis: A new measure of patient-physician communication. *Social Science & Medicine (1982)*, 198, 22–26. https://doi.org/10.1016/j.socscimed.2017.12.021

Wulff, D., U., Hills, T., T., & Mata, R. (2022). Structural differences in the semantic networks of younger and older adults. *Scientific Reports*, 12, 21459. https://doi.org/10.1038/s41598-022-11698-4

Xiong, M., J., Logan, G., D., & Franks, J., J. (2006). Testing the semantic differential as a model of task processes with the implicit association test. *Memory & Cognition*, 34(7), 1452–1463. https://doi.org/10.3758/bf03195910

Yarosh, D., B. (2019). Perception and deception: Human beauty and the brain. *Behavioral Sciences (Basel, Switzerland)*, 9(4), 34. https://doi.org/10.3390/bs9040034

Zeki, S., Chén, O., Y., & Romaya, J., P. (2018). The biological basis of mathematical beauty. *Frontiers in Human Neuroscience*, 12, 467. https://doi.org/10.3389/fnhum.2018.00467

Considering the emotion of disgust in the context of terminology and contemporary literature

Renáta Kišoňová

Abstract. This chapter focuses on the terminological clarification of moral and social emotions of disgust, its synonyms and antonyms, and contextual consideration of this emotion (evolutionary, biological, social context). The aim of the submitted text is a more detailed mapping of the concept of disgust both through literary and philosophical classics (Jean-Paul Sartre) and through contemporary novellas (Hans Tallis, Han Kang). Magnetizing dealing with disgust is not a rarefied taste; popular fiction like horror or detective stories as well as literary novels is filled with interesting uses of disgust. I am interested in how the concept of disgust is used and understood in contemporary literature works and how is the term used in different cultures and terminology (Dutch, Burkina Faso, England).

Introduction

Disgust is a multifaceted emotion covering a number of areas: food, disease, sexual manners, social status, environment, condemnation and injustice. Within biological accounts, those that give disgust its universal visceral meaning, this emotion is seen as a danger response traced to an organism's preservation (Ryynanän, Kosonen, Ylönen 2023, 4). As Paul Rozin (Rozin, Singh 1999) or Jonathan Haidt (2003, 2013) reflect, this danger function does ring true when we think of the bodily recoil related to harmful foods and infectious diseases. Valt Curtis (2013) understands dangerous foods, faeces and sexual activities dealing with bodily excreta, diseased-looking and dying humans, and dead carcasses of animals as instances that are related to disgust's function against shielding humans from disease and death. Every flu season, sneezing or coughing become the day-to-day background noise in every workplace, and coworkers tend to move as far - and as quickly - away from the source of these bodily eruptions as possible. It seems that humans recoil from objects that they perceive as dirty and even struggle to overcome feelings of discomfort once the offending item has been cleaned. These reactions are universal, and although there are cultural and individual variations, by and large, we are all disgusted by the same things (Curtis 2013). *Disgust* belongs to the seven universal emotions and arises as a feeling of aversion towards something offensive. We can feel disgusted by something, or someone we perceive with our physical senses, by the actions or appearances of objects, people and surprisingly even by ideas (Ekman 2003, 66-68). Disgust contains a range of states with varying intensities from mild dislike to intense loathing. All states of disgust are triggered by the feeling that something is aversive, repulsive and/or toxic. The universal trigger for disgust is the feeling that something is offensive, poisonous or contaminating. Some triggers for disgust are universal (such as encountering certain bodily products), whereas other triggers are much more culturally and individually influenced (such as certain types of food) (Ekman 2003, 66-68).

General Definition of Disgust

Moral philosophy and moral psychology (as well as other disciplines focusing on morality) had very little to say about emotion of disgust. For example, Aurel Kolnai focused on disgust in a long essay in 1929 called *On Disgust* as a rare exception. He mentioned "the problem of disgust has to my knowledge been thus far sorely neglected... disgust – although a common and important element of our emotional life – is a hitherto unexplored sphere. At best it has been occasionally dis-cussed as a 'higher degree of dislike', as 'nausea', or as 'reaction following a repression of urges'" (Kolnai, 2004, 29). In this sense Kolnai urges on a feature of disgust that might account for its previous marginalisation. In its fundamental form, it almost equally qualifies as a type of sensory reaction, rather than an emotion, perhaps rather like *the startle reflex*. According to him, disgusting objects taunt us and press upon us in spite of our initial recoil. In Kolnai´s description, emotion of disgust exhibits "shameless and unrestrained forcing itself upon us. The disgusting object grins and smirks and stinks menacingly at us" (Kolnai, 2004, 41). As per Kolnai's understanding, disgust is an eminently aesthetic emotion and disgusting objects exert a grisly attraction. The disgusting has an allure it exerts a fascination which manifests itself in the difficulty of averting our eyes at a gory accident, for example, of not checking out the quantity and quality of our excretions, or in the attraction of popular horror films (Miller 1997). Nina Strohminger and Victor Kumar, editors of *The Moral Psychology of Disgust*, mentioned in the "Introduction" to the book one reason of the fact that emotion of disgust is neglected in philosophical work. They see a reason in a general antisentimentalist bent, in which emotions are imprecise and resistant to measurement (Strohminger, Kumar, 2020, 1). And mainly, "disgust seemed more like a drive than an emotion-the counterpart to appetite, perhaps only a notch above nausea, and therefore not of much concern to the self-respecting psychologist" (Strohminger, Kumar, 2020, 1). Aristotle presented the basic emotions of anger, love, hate, meekness, fear, courage, shame, kindness, compassion, envy and rivalry in the second book of Rhetoric in relation to the speaker's personality. He does not interpret the emotion of disgust, nor does he use this term in Rhetoric; he interprets emotion of *cataphronesis*, contempt, which may be the closest to social emotion of disgust. He does not also interpret the emotion of admiration (as an antonym of disgust), nor does he use this term in Rhetoric; he interprets the so-called *zelos* (English uses the term "emulation" for the word *zelos*), i.e. rivalry and its opposite, cataphronesis, contempt. However, of all the emotions to which Aristotle[3] paid attention, it is zeal and rivalry that is closest to the emotions of admiration and to what is both its presupposition and result, to authority.

When we move to contemporary understanding of emotions in philosophy, we come across many concepts and approaches. For example, Démuth, Démuthová and Keceli understand emotions as "a complex and psychosomatic reaction that includes several, sometimes even highly distinct and contradictory, aspects" (Démuth, Démuthová, Keceli 2022, 112), or on the other place Andrej Démuth understands emotions as "...multidimensional objects which form continuously divided space, in which it would be possible to differentiate individual units of meaning well on the basis of different saturations of individual dimensions" (Démuth, 2019,

---

[3] With this consideration, Aristotle began an idea that was later elaborated by social psychology in connection with the concept of the so-called identification (e.g., imitating a father by a son), according to which an individual does not imitate another subject because it is a source of satisfaction for him but because he behaves like an admired person and thereby increases his own self-esteem (Nakonečný, 2009).

19). These definitions can also include the emotion of disgust which also appears to be multilayered, directed towards physicality as well as sociality and morality.

I will take a closer look at three books in which the emotion of disgust (or nausea) is dealt with. First of all I have chosen Sartre's novel *Nausea*, where he deals with nausea as feelings of the futility of human life ad the loss of the meaning of being.

Novel *Nausea*

Jean-Paul Sartre has described in his book *Nausea* deep and existential experience of disgust, which has been gushing from his main character's feelings of persons, things around him, his own face and body. *Nausea,* one of the canonical works of existentialism, is the story of Antoine Roquentin, a writer who is horrified at his own existence. In impressionistic, diary form he ruthlessly catalogues his feelings and experiences. His thoughts culminate in a pervasive, overpowering feeling of nausea which "spreads at the bottom of the viscous puddle, at the bottom of our time – the time of purple suspenders and broken chair seats; it is made of wide, soft instants, spreading at the edge, like an oil stain" (Sartre, 1964, 21). Sartre used the term *nausea*, which is used as one of many synonyms for disgust in English. Etymologically appeared the notion nausea in the early fifteenth century in the meaning vomiting, from Latin nausea, which is related to seasickness, from Ionic Greek nausia, which can be translated seasickness, nausea, disgust, literally ship-sickness, from naus ship. Despite its etymology, the word "nausea" in English seems never to have been restricted to seasickness (Harper, 2020). The sixteenth-century canting slang had *nase*, or *nasy* which meant hopelessly drunk. The *nāu-*, has Proto-Indo-European root, and it refers to "boat", "to a sickening extent". Latin, literally means "to sickness", "from ad to" and "nauseam", accusative of nausea, especially of the disgust aroused by wearisome repetition.[4]

The Thesaurus lists these terms as synonyms: airsickness, biliousness, car sickness, mal de mer, motion sickness, nauseousness, qualms, queasiness, regurgitation, retching, vomiting, abhorrence, aversion and finally disgust, distaste and so on. Antonyms for nausea are these terms: relish, fondness, liking, partiality predilection, etc. ("Nauzea", Dictionary.com 2022).

Roquentin's predicament is not simply depression or simply anxiety, although his experience has pushed him to that point. Sartre presents Roquentin's difficulties as arising from man's inherent existential condition. Nausea is disgust at the traumatic decomposition of the divine within existence, symptomatic of the discovery of the absurd, of the disenchantment of the world. Transcendence and providence were invented by man. Every being is meaningless in itself. "Things are bad! Things are very bad: I have it, the filth, the Nausea. And this time it is new: it caught me in a cafe. Until now cafes were my only refuge because they were full of people and well lighted: now there won't even be that any more; when I am run to earth in my room, I shan't know where to go" (Sartre, 1964, 12).

Disgust means in this context not only experience and feeling but existential necessity how to process absurdity and senselessness of being. When watching his own face in the mirror, Roquetin reflects: "I see a slight tremor, I see the insipid flesh blossoming and palpitating with abandon. The eyes especially are horrible seen so close. They are glassy, soft, blind, red-rimmed, they look like fish scales. I lean all my weight on the porcelain ledge, I draw my face closer until it touches the mirror. The eyes, nose and mouth disappear: nothing human is left. Brown wrinkles show on each side of the feverish swelled lips, crevices, mole holes. A silky white down covers the great slopes of the cheeks, two hairs protrude from the nostrils: it is a geological embossed map. And, in spite of everything, this lunar world is familiar to me" (Sartre, 1964, 13). The biological dimension of nausea was expressed by Sartre also with the words: "I wanted to vomit. And since that time, the Nausea has not left me, it holds me" (Sartre, 1964, 14). The reader can reflect Sartre´s terminology related to physical body but also leading to the moral and social experience of disgust (nausea).

Novel *The Vegetarian*

Another example of literature working with the concept of emotion of disgust is a novel by Korean author Han Kang[4] *The Vegetarian*. *The Vegetarian* is fable-like in structure and works with the term "disgust" in English translation. It centres on the vivid self-destruction of a single human body. That body belongs to a housewife named Yeong-hye, who is described by her husband as "completely unremarkable in every way" (Kang, 2015, 10) and "she made for a completely ordinary wife who went about things without any distasteful frivolousness" (Kang, 2015, 10). For Mr. Cheong, who has "always inclined to the middle course in life" (Kang, 2015 10), this is part of her appeal.

"The passive personality of this woman in whom I could detect neither freshness nor charm, or anything especially refined, suited me down to the ground" (Kang, 2015). Yeong-hye's husband says. Yeong-hye's decision not to eat meat is received as an appalling rebuke by her entire family, especially her father, a Vietnam War veteran whose violent tendencies suggest the traumas of the battlefield. During a family meal, orchestrated as an intervention of sorts, he attempts to shove a piece of sweet-and-sour pork down his daughter's throat. In response, Yeong-hye slits her wrist as the entire family watches in horror. Finally, she is institutionalised as a result of social and surrounding misunderstanding which she is experiencing. According to reviewers it's a bracing, visceral, system-shocking addition to the "Anglophone reader's diet". It is sensual, provocative and violent, ripe with potent images, startling colours and disturbing questions (Hahn 2015).

Disgust of meat and actually most foods succeeded delineate very authentically, physically, sensually. "Along bamboo sticks trung with great blood-red gashes of meat, blood still dripping down. Try to push past but the meat, there's no end to the meat,

---

4 Han Kang was born in Gwangju in 1970. She won the 25th Korean Novel Award for the novel *Baby Buddha* in 1999, was awarded Manhae Literary Prize for Human Acts in 2014, and the novel *The Vegetarian* won the 2016 Man Booker International Prize. Han Kang currently works as a professor in the Department of Creative Writing at the Seoul Institute of the Arts.

and no exit. Blood in my mouth, blood-soaked clothes sucked onto my skin" (Kang, 2015, 14). An interesting research related to disgust and food made the team investigate visual attention to disgusting oral, non-oral and control non-disgusting stimuli using a touch-screen paradigm with a sample of 60 adult participants in Slovakia (Fančovičová, Prokop, Šramelová, Gaëtan, Méot, Witt, Bonin, Medina-Jerez, 2021). They found that disgusting pictures triggered visual attention more than control pictures, and although participants identified disgusting food items quicker than non-disgusting food pictures, differences did not reach statistical significance. Research study concludes that "the evolution of disgust could have been originally favoured by the repulsion of contaminated food, but the benefits from disease avoidance were soon extended to disgust sensitivity to pathogens that threaten our bodies using non-oral entry points" (Fančovičová, Prokop, Šramelová, Gaëtan, Méot, Witt, Bonin, Medina-Jerez, 2021, 2). In the novel *The Vegetarian*, the reader may literally feel, experience and undergo disgusting perceptions which are also reflected on his or her face. "..it's the smell." "The smell?" "The meat smell. Your body smells of meat." "Didn't you see me just take a shower? So where's this smell coming from, huh?" "From the same place your sweat comes from," she answered, completely in earnest. "This was just too ridiculous forwords" (Kang, 2015, 17). The author of *The Vegetarian* uses in the story emotion of disgust related to human sense of smell in the first plane and in others the heroine also experiences disgust on a visual and haptic level. The other characters experience disgust more morally; they condemn the heroine Yeong-hye. She suffers whole story with prejudices of her vicinity who cannot (and not even try) to understand her aversion to meat. She is condemned to loneliness and misunderstanding because of her otherness. Disgust is emotion which experiences Yeon-hye, but it is also moral and social disgust which experience her family regarding her behaviour and her decision not to eat meat. I suppose that also prejudice is present in this novel on both the sides: Yeong-hye experiences prejudices related to her vicinity and vice versa. As she is changing in the story, the language shifts, too, moving between the baffled irritation of Mr Cheong's first-person narration in part one, the measured prose of Yeong-hye's father world, the dense and bloody narrative of Yeong-hye's dreams, and seductive descriptions of living bodies painted with flowers – in states of transformation or wasting away (Hahn, 2015). And this experience of mutual disgust is also experienced by the reader very emotionally. *The Vegetarian* story is disturbing, troubling and disquieting, forcing the reader to question the very nature of human life and character. In this sense, Han Kang uses emotion of disgust truly spellbinding and unputdownable. The author is gripping and dismaying at the same time, mixing sheer aesthetic beauty and utter psychological and physical distress still working with an object that disgusts either physically or morally. In this sense is Spinoza's claim "for a man at the mercy of his emotions is not his own master but is subject to fortune, in whose power he so lies that he is often compelled, though he sees the better course, to pursue the worse" (Spinoza, 1982, 153) eloquent and truthful[5].

Evolutionary Consideration Related to *The Vegetarian* Novel

Despite being perfectly nutritious, consuming bugs is considered gross in many cultures including the West (Rottman, DeJesus, Gerdin, 2020, 27). What is the function of irrational behaviour? Do we experience disgust towards bugs because of disease risk? (Dunbar, 1996) Disgust evolved because it has helped us to avoid physical contact with parasites and poisons. "Disgust also regulates social interactions outside the moral domain. This is particularly evident for eating behaviors, as food is steeped in social connotations" (Rottman, DeJesus, Gerdin, 2020, 36). Some animals are considered notoriously unpopular among people (not meaning as a food). Using the forced-choice paradigm, researchers in Slovakia investigated which cues humans perceive as disgusting in spiders, one of the most unpleasant animals in the world (Zvaríková, Prokop, Ježová, Medina-Jerez, Fedor, 2021), and they found that perceived disgust of spiders was triggered by enlarged chelicerae,

---

5 Spinoza's *Ethics* is permeated with an extensive catalogue of diferent emotions, mentions disgust only once and in passing.

enlarged abdomen and the presence of body hair (Zvaríková, Prokop, Ježová, Medina-Jerez, Fedor, 2021). Female respondents considered hairiness more disgusting than males. Body hair correlates with disgust sensitivity, reflecting Prokop's hypothesis (Prokop et al., 2013). It means that hairy bodies can suffer from high loads of ectoparasites that end up transferring diseases to the host animal and ultimately decreasing the fitness of an individual (Prokop et al., 2013). Evolutionary theorists who have searched for the origins of human morality have generally found its source in the dynamics and difficulties of reciprocal altruism. Like Jonatan Haidt mentioned, many social species, from vampire bats to chimpanzees, have figured out the trick of playing tit-for-tat within dyads such that cooperating pairs end up reaping more benefits than either member would on their own (Haidt, 2003). Disgust is a response to physical objects and social violations. For example, Lazarus resorted united the physical and social aspects of disgust: "[T]aking in or standing too close to metaphorically speaking an indigestible object or idea" (Lazarus, 1991). Haidt argues that "disgust grew out of a distaste response found in other animals, which was then shaped by evolution to become a more generalized guardian of the mouth. Disgust rejects foods not principally for their sensory properties but for their ideational properties (e.g., the source of the food, or its contact history)" (Haidt 2003). According to Haidt, in many cultures and languages, the words and facial expressions used to express disgust towards rotting meat or faeces are also used to condemn social transgressions that do not involve the body in any physically disgusting way (Haidt 1997). For example, for Westerners, sociomoral disgust can be described most succinctly as the guardian of the lower boundary of the category of humanity. People who "degrade" themselves, or who blur the boundaries between humanity and animality, elicit disgust in others (Haidt, 2003). The action tendency of disgust is according to Haidt often prosocial. For those who trigger moral disgust, people in a society set up a reward and punishment structure that acts as a strong deterrent to culturally inappropriate behaviours, particularly those involving the body (transgender people, gays, lesbian, etc.) (Haidt, 2003). Emotions can influence moral cognition (Trivers, 1971). There are experimental studies that repugnance is typically an effect of moral judgement, not a cause (May, 2020, 156). "Eating an insect might disgust you, but it needn't have any relation to your moral beliefs, either as an effect or as a cause. However, when there is a connection between a moral belief and repugnance, the emotion may be elicited by the belief, not the other way around" (May, 2020, 156). The universal function of disgust is to get away from, block off or eliminate something offensive, toxic or contaminating (Ekman, Wallace, 2003). "One evolutionary benefit of disgust is to keep us away from or remove things potentially dangerous or damaging to keep us safe and healthy (e.g., not eating something putrid, staying away from open sores to avoid catching an infection or disease, avoiding interactions with "morally tainted" people)" (Ekman, Wallace, 2003, 66–68). Disgust can be also dangerous (Ekman, 1992). Many societies teach the avoidance of certain groups of people deemed physically or morally disgusting and, thus, can be a driving force in dehumanising and degrading others, which may reach to superiority, prejudicies and other social deviations. On the other hand, disgust can be suppressed, while witnessing gross bodily functions like bleeding, defecating, vomiting, etc. in others often evokes disgust, this reaction is suspended when it is someone with whom we are related, or close. (Kišoňová, 2021, 24) "Intimacy lowers the threshold for what we consider disgusting. So, while we still may feel some degree of disgust, it is reduced enough that we are able to help those we care about. Now, rather than try to get away, we are called to reduce the suffering of the loved one (e.g., changing a baby's diaper or taking care of a sick family member). This suspension of disgust establishes intimacy and may even strengthen love and community" (Ekman, Wallace, 2003, 66–68). Not every vegetarian, for example, becomes disgusted by meat, but for those who have moral reasons, there is empirical evidence that they are more disgusted by meat than those who are vegetarians for health benefits. The emotional response is here related to the moral judgement by following it (May, 2020, 157).

The Novel *Vienna Secrets*

Another example of a novel that manipulates disgust to serious end is series of mysteries set in the early years of the twentieth century in Freud's Vienna, from Frank Tallis, *A Death in Vienna, Vienna Blood, Fatal Lies* and *Vienna Secrets*. I have chosen this book series in my mapping of disgust terminology and broader context because this book shows, on one hand, wide terminology reflecting disgust and disgusting physical body parts and, on the other hand, focuses on the relation among disgust and prejudices. The series presents a vivid picture of the city and its complex cultural organisation at that period of history: Vienna's long-standing anti-Semitism is on the rise, and the reader realises how vicious it will become in reality. The reader is immersed in a fictional world perceived with a sort of double-vision, because the main characters love their city and believe its cultural virtues override the political and social discomforts that they acknowledge are growing around them. And again, the reader realises better (Korsmeyer, 2022). The main character is a doctor Lieberman, a secular Jew who applies psychoanalytic methods to assist the police in solving crimes. Those are horrendous. The reader is introduced to mutilated corpses in despicable poses, set in areas of the city freighted with meaning, and described in sickening detail. Constables called to the scene often have to vomit at what they see. The horror of an event indicated by vomiting on the part of novice policemen is not an uncommon plot device in the story (Korsmeyer, 2022; Tallis, 2009). "It foregrounds a diference between giving into disgust and overcoming its power, for seasoned ofcers – especially those who are our main protagonists – are appalled and repulsed, but their dignity is rarely compromised by the loss of control that vomiting indicates. Giving into extreme disgust relinquishes power over one's own body, embarrassing and undignifed, even if perfectly under-standable, for even reading about the abominations is challenging. Just how do the more experienced ofcers overcome what would be a natural disgust response? Rarely do they fail to respond at all; the emotion is not simply suppressed." I used longer quotation by Tallis's interpreter Korsmeyer, who used many synonyms of disgust in his reading of *Vienna Secrets* set (vomiting, appalled, repulsed, abominations). And now let the Tallis series character himself be read and let the individual examples of what causes disgust be read: "Liebermann advanced and made his way – somewhat warily – around the expanse of congealed blood. He squatted and looked directly into the truncated stump of the monk's neck.

The dawn sky provided him with just enough light to identify the remains of the key cervical structures; however, what he observed was nothing like the cross sections that he remembered from his anatomy classes, which had resembled the fatty marbled meat of a freshly sliced joint. The aperture of the trachea was displaced, as were the hardened remnants of cartilage. The vertebrae were fractured, and the muscles ripped and twisted. A rubbery length of artery hung out over the trapezius, still dripping. Something purple, veined, and lobu-lated was lying on the ground close to the monk's right shoulder. Liebermann guessed that it might be a piece of the thyroid gland... He stood up and moved toward the severed head. It seemed to take him an inordinate amount of time to travel the relatively short distance – and all the while the horrifc object exercised a curious fascination" (Tallis, 2009, 5–6). Tallis showed interesting movement from physical disgust (vomit, blood, multilated corpses, etc.) to social and moral disgust from human group behaviour, customs, religion and living. I have to mention interesting current research, from 2021 (published in March 2022), investigated data from adult heterosexuals from 31 countries. Research showed a small relation between pathogen disgust sensitivity and measures of antigay attitudes (Leeuwen et al., 2022). "Analyses showed that pathogen disgust sensitivity relates not only to antipathy toward gay men and lesbians, but also to negativity toward other groups, in particular those associated with violations of traditional sexual norms (e.g., prostitutes)" (van Leeuwen, et al., 2022, 2). This research shows that the association between pathogen-avoidance motivations and antigay attitudes is relatively stable across cultures and is a manifestation of a more general relation between pathogen-avoidance motivations and prejudice towards groups associated with sexual norm violations (van Leeuwen et al., 2022); in other words, it showed that the disgust may lead to social prejudices.

The Term "Disgust" and Its Etymology

I will focus now on the term *disgust*, its etymology, synonyms and antonyms. Oxford Compact English Dictionary defines disgust as "strong revulsion or profound indignation" (Soanes, 2003, 311). The definition describes and indicates both physical manifestations of disgust and moral and social aspects (indignation is directed towards morally pointing out our disagreement, displeasure with something or someone). The word "disgust" came into English in 1601 from the Old French word "desgouster" meaning distaste, loathe or dislike, in the sense of giving a bad taste to one's mouth and from Latin "des (dis) gustus", which means tasting. The word "gustus" was very common from the beginning of the nineteenth century from Italian gusto which can be translated as "taste", from Latin "gustus", "a tasting", related to "gustare", "to taste", take a little of". English first borrowed the French form, guste "organ of taste; sense of taste" (mid-fifteenth century), but this became according to Etymology Dictionary obsolete (Harper, 2020). The Proto-Slavic *gnus* is related to the Greek *chanuo*, which can be translated as pinching off, while the noun *chnoos* is translated as *foam*, sediment, mud and dust. The meaning of the expression apparently developed in the direction of "that which is rubbed off", broken and rotten, that which arouses resistance. The Thesaurus lists as synonyms of the term "disgust" following words: antipathy, aversion, dislike, distaste, hatred, loathing, repulsion, revulsion, abhorrence, abomintion, nausea, etc. Antonyms of disgust are approval, liking, love, admiration, appeal, desire, respect, etc. I work mainly with the term "admiration" as opposite to the term "disgust" in this text.

Emotion of Admiration as One of the Antonyms of Disgust

Admiration belongs to positive emotions in response to an outstanding person or object. This emotion should serve to keep a person's ideals and values accessible as guides for behaviour and also contribute to the adoption and internalisation of ideals, values and goals (Schindler, Zink, Windrich, Menninghaus, 2013, 85–118). As mentioned in Schindler, "...admiration is elicited by outstanding role models who represent specific ideals or values. The excellence of such models, at least in principle, can be understood, matched, and even surpassed by others" (Schindler, Zink, Windrich, Menninghaus, 2013, 85–118). Moral function of admiration may be seen in encouraging others who aspire to grow by showing that it is possible to actualise ideals. The action tendencies associated with admiration are to uphold and honour ideals. The admiring one seeks to praise and affiliate with the other as well as to emulate the other's conduct. The primary function of admiration is to enhance the individual's agency in striving for ideals (Schindler, Zink, Windrich, Menninghaus, 2013).

Etymology of Admiration

The word "admiration" is in English the noun of the verb admire, which is defined as "regard with respect or warm approval; look at with pleasure" (Soanes, 2003,13). Etymology of the word "admiration" reaches into the early fifteenth century, from Old French admiration which means "astonishment, surprise" and from Latin admirationem a wondering at, admiration, noun of state from the past-participle stem of admirari "regard with wonder, be astonished", from ad "to; with regard to" and mirari "to wonder", from mirus which may be translated as "wonderful". The sense has gradually weakened since the sixteenth century towards high regard, "esteem".

Cultural and Linguistic Approach to Disgust

Disgust is a response to physical objects, to social violations, and social prejudice may be one of the defence mechanisms. "Eating an insect might disgust you, but it needn´t have any relation to your moral beliefs, either as an effect or as a cause. However, when there is a connection between a moral belief and repugnance, the emotion may be elicited by the belief, not the other way around" (May, 2020, 156). The most discussed and most researched area of emotions includes the problem of cultural conditionality. For example, as mentioned by Parkinson, Fischer, Manstead (2005) – English language contains approximately 2,000 words that express emotions, Dutch around 1,500, Taiwanese Chinese about 750 and Malay 230.

I will refer to Valerie Curtis research, in which she investigated understanding of disgust in various languages and cultures (Dutch, English and Burkina Faso). When women in England, Burkina Faso, India and Netherlands wrote or spoke about disgust in Curtis´s research, they most frequently refer to disgusting things in the kitchen (the question was, why they clean, why they use soap and detergents and so on) as: faeces. Dutch women found revolting except faeces of cats, aphids in lettuce, hairs, dogs pollution, vermin, vomit, worms, flies and so on (Curtis, 2013, 2). Women in Burkina Faso found disgusting except faeces, dirty latrine, dirty food, unswept yard, worms, sexual relations before child is weaned, smelly drain, dirty clothes, vomit, sick people and so on. English women refer to stained kitchen, dirty hotel, dog shit, flies, drunks, vomit, rude people and dirty nails (Curtis, 2013). The cultural and linguistic differences did not represent a significant determinant; on the contrary, there are many overlaps between the answers. We can see not only physically meanings of disgust (worms, vomit, sickness, etc.) but moral and social as well (sexual relations before child is weaned f.e.). When Curtis made similar research covering more cultures (literary whole world, because she questioned people at international departure lounge of Athens Airport) the answers referring to disgust were: dog faeces in the street, faeces, dirty baby diapers, animal saliva, people who do not brush their teeth, polititians, rude Europeans, all Americans, materialism, injustice, dying person and so on (Curtis, 2013). And again, disgust has moral, social and biological or physical contexts in respondents' answers.

Conclusion

"But the more I watched, the more I wanted to learn, and unlike Rosa I became puzzled, then increasingly fascinated by the more mysterious emotions passers-by would display in front of us. I realized that if I didn't understand at least some of these mysterious things, then when the time came, I'd never be able to help my child as well as I should. So I began to seek out – on the sidewalks, inside the passing taxis, amidst the crowds waiting at the crossing – the sort of behavior about which I needed to learn... " (Ishiguro, 2021, 20) said Klara, the AI in novel *Klara and the Sun*, by Nobel Prize winner, Kazuo Ishiguro. To understand disgust means to understand its terminology, evolutionary context, its etymology, and moral and social contingency. I have tried to focus on some interesting results of current research in this text and to indicate the meanings of the emotion of disgust for human morality and sociality.

References

Curtis, V. (2013). *Don´t Look, Don't Touch, Don't Eat*. Chicago: University of Chicago Press.

Démuth, A. (2019). *Beauty, Aesthetic Experience, and Emotional Affective States*. Berlin: Peter Lang Verlag.

Démuth, A., Démuthová, S. Keceli, Y. (2022). A semantic analysis of the concept of anger and its connotations (based on free associations experiment) in the Turkish language. *Luboslovie*, 22, 111–127.

Dunbar, R. (1996): *Grooming, Gossip, and the Evolution of Language*. Cambridge, MA: Harvard University Press.

Ekman, P. (1992). An argument for basic emotions. *Cognition and Emotion*. 6 (3/4), 169–200.

Ekman P., Wallace V. F. (2003). *Unmasking the Face: A Guide to Recognizing Emotions from Facial Clues*. Los Altos, (CA): Malor Books.

Fančovičová, J., Prokop, P., Šramelová, D., Gaëtan, T., Méot, A., Witt, A., Bonin, P., Medina Jerez, W. (2021). Does food play a prominent role in visual attention to disgusting stimuli? *Journal of Ethology*, 40 (1). https://doi.org/10.1007/s10164-021-00722-1.

Hahn, D. (2015). *The Vegetarian by Han Kang Review – an Extraordinary Story of Family Fallout*. In: The Guardian, 24.1. https://www.theguardian.com/books/2015/jan/24/the-vegetarian-by-han-kang-review-family-fallout.

Haidt, J. (2013). *The Righteous Mind*. New York: Penguin books.

Haidt, J. (2003). *Moral Emotions*. Oxford: Oxford University Press.

Harper, D. (2020). "Disgust" In: *Online Etymology Dictionary*. Retrieved from: https://www.etymonline.com/search?q=disgust.

Ishiguro, K. (2021). *Klara and the Sun*. New York: Alfred A. Knopf.

Kang, H. (2015): *Vegetarian*. London: Hogarth.

Kišoňová, R. (2021). The social emotions of disgust and admiration in the context of language. *Language, Individual and Society*, 15, 22–27.

Kolnai, A. (2004). *On Disgust*. Edited and an introduction by B. Smith and C. Ksorsmeyer. Chicago: Open Court.

Korsmeyer, C. (2022). Overcoming disgust: Why, when and whether. In: *Cultural Approaches to Disgust and Visceral*. New York: Routledge.

Lazarus, R. S. (1991). *Emotion and Adaptation*. Oxford: Oxford University Press.

May, J. (2020). *The Limits of Appealing to Disgust*. Lanham: Rowman, & Littlefield International Ltd.

Miller, W. I. (1997). *The Anatomy of Disgust*. Cambridge, MA: Harvard University Press.

Nakonečný, M. (2009). *Sociální psychologie*. Praha: Academia.

Parkinson B., Fischer A., Manstead A. S. R. (2005). *Emotion in Social Relations: Cultural, Group, and Interpersonal Processes*. Philadelphia, PA: Psychology Press.

Prokop, P., Rantala, M. J., Usak, M., Senay, I. (2013). Is a woman's preference for chest hair in men influenced by parasite threat? *Archives of Sexual Behavior*, 42, 1181–1189.

Rottman, J., DeJesus, J., Gerdin, E. (2020). The Social Origins of Disgust. In: Strohminger, N., Kumar (ed.), *The Moral Psychology of Disgust*. Lanham: Rowman, & Littlefield International Ltd.

Rozin, P., Singh, L. (1999). The moralization of cigarette smoking in the United States. *Journal of Consumer Psychology*, 8 (3), 321–37.

Ryynänen, M., Kosonen, H. S., Ylönen, S. C. (2023). *Cultural Approaches to Disgust and the Visceral*. New York: Routledge.

(s. a.) (2022): "Nauzea". In: *Dictionary.com*. Retrieved from https://www.thesaurus.com/browse/nausea.

Sartre, J. P. (1964). *Nausea*. New York: New Directions Publishing.

Schindler, I., Zink, V., Windrich, J., Menninghaus, W. (2013). Admiration and adoration: Their different ways of showing and shaping who we are. *Cognition & Emotion*, 27(1), 85–118. https://doi.org/10.1080/02699931.2012.698253.

Soanes, C. (2003). *The Oxford Compact English Dictionary*. Oxford University Press.

Spinoza, B. B. (1982). *The Ethics and Selected Letters*. Indianopolis, (IN): Hackett Pub. Co.

Strohminger, N., Kumar, V. (2020). Introduction: Disgust: A cross-pollination. In: Strohminger, N., Kumar (ed.). *The Moral Psychology of Disgust*, London: Rowman, & Littlefield International Ltd.

Tallis, F. (2009). *Vienna Secrets*. New York: Random House.

Trivers, R. L. (1971). The evolution of reciprocal altruism. *Quarterly Review of Biology*, 46, 35–57.

van Leeuwen, F., Inbar, Y., Petersen, M. B., Aarøe, L., Barclay, P., Barlow, F. K., de Barra, M., Becker, D. V., Borovoi, L., Choi, J., Consedine, N. S., Conway, J. R., Conway, P., Adoric, V. C., Demirci, E., Fernández, A. M., Ferreira, D. C. S., Ishii, K., Jakšić, I., ... Tybur, J. M. (2022). Disgust sensitivity relates to attitudes toward gay men and lesbian women across 31 nations. *Group Processes & Intergroup Relations*, 0(0). https://doi.org/10.1177/13684302211067151

Zvaríková, M., Prokop, P.,Ježová, Z, Medina-Jerez, W., Fedor, P. (2021). What Makes Spiders Frightening and Disgusting to People? *Frontiers in Ecology and Evolution*, 9, 694569. https://doi.org/10.3389/fevo.2021.694569.

Spiritual and Theological Discernment of Good and Evil

Ľubomír Batka

Abstract. Discernment between good and evil is a judgement between opposites. There are different realms where discernment of good and evil plays a major role, defined by a concrete authoritative standard (law, religion, medicine).

Discernment of good and evil in the religious sphere can be divided in spiritual realm dealing with the decisive question of who has the right to discern. The *discretio spiritum* was in the history of the Christian Church ascribed either to theologians or to the clerical offices in the Church or to the monastic way of life.

In the theological realm, the discernment of good and evil is related to the question of origin of evil. The decisive role plays the argument about the condition of the person making the discernment. Sin and grace/faith are theological categories leading to a conclusion that a sinner cannot discern properly between good and evil. The discernment in the state of sin is dependent on emotions like remorse or feeling of guilt. In the state of grace, the discernment depends on faith.

Key Biblical passages and concepts of Augustine, Bernhard of Clairvaux, Ignatius of Loyola, and Martin Luther are presented in two chapters. The closing chapter deals with the question if Artificial Intelligence would be capable of discernment of good and evil rather in the spiritual or theological sense.

Introduction

The discernment of good and evil is a capability of human beings. Regardless of how the medium of discernment is described – as reason, conscience, emotions, *Urteilskraft*, *raisonnement*, remorse, penance, inner reflection, *diakrisis* (Rom. 12:2; 1 Thess. 5:19-22; Eph. 5:10; Phil. 1:10; Heb. 5:14) – always a human subject is involved. The Greek term *diakrinein* as well as the Latin *discernere* point to the activity of making a distinction by a person. Only a subject is able to discern between good and evil. The process of change, involving the will of a subject, is called *metanoia*, revival, awakening, rebirth or healing.

For further inquiry in this text, we will rely on conceptual analysis of terms "good" and "evil", in order to show and describe relations between terms in a particular conceptual system, as proposed by Zouhar (Zouhar, 2016). We attempt to find terms that identify good and evil in religious realm in the same meaning, but the modus of identification of the content is better or more transparent.

Discernment between broad categories of good and evil means a discernment between opposites. Generally, good is worth pursuing, and evil is to be avoided. Evil is associated with negativity, harm, destruction, brokenness, nothingness and death. Good is associated with perception of wholeness, progress, beauty, harmony, fullness and life. The aim of discerning is to be able to make a right choice (*eligere*) and to achieve a good, just, moral, healthy life and to avoid evil, unjust, immoral and unhealthy life. Such contradictory and opposing distinctions intend to process common experiences but express them in different ways.

The content of good and evil depends on which aspect of life is concerned – the spiritual, legal, moral, theological or empirical. Here, various categories of discernment can be used. For example, in the realm of legality, categories of just and unjust are used. In the realm of morality, one speaks about the moral and the immoral, in the realm of metaphysics about the natural and the non-natural, and in the realm of medicine about health and disease.

Distinct realms of discernment are created, and in certain cases even defined, by an authoritative standard (Law, Ethical system, Philosophical preconceptions, Canon Law, Bible, Word of God, Confessions). For example, the primary normative standard for a Protestant theology is the Bible (*norma normans*), and in some cases the secondary normative standard is the Confessions (*norma normata*). Discussion appears around the decisive authority in discernment and about the interpretation of the chosen authoritative standards. Discernment is therefore connected to a process of evaluation among multiple authorities as well as to the discussion about who has the ultimate right of authoritative interpretation.

In this chapter we concentrate on the realm of Christian religion. The main concern of religion can be described as a proper attitude to God that leads to a good quality of life: salvation, joy, peace, consolation fruits of the Spirit, love (Weismayer, 1983). In the realm of religion, the discernment can be subdivided into two areas: spiritual and theological.

The goal of spiritual discernment is to decide which spiritual experience is true and which one is false. Hence the question arises about who has the right to discern. In the realm of theology, the origin and nature of evil are relevant. Followingly the nature of the person making the discernment is of main concern. Whereas in the state of sin, emotions like feelings of guilt are relevant, in the state of grace. the category of faith becomes predominant. In theological realm the discernment is related to the question of whether a sinful person can discern properly.

Spiritual Discernment

In the Bible, distinguishing true and false prophets play an eminent role especially in the Old Testament (cf. Dt. 13:2-6; Dt. 18:20-22; 1 Sam. 16:14; Jer. 5:31; Ezek. 22: 28) and has a crucial importance for the survival of Israel. In the New Testament, a false prophecy has even an apocalyptic character (1 Tim. 4:1; Mk. 13; Mt. 24, Lk. 21). It is characterised with deceit and is related to evil (Rev. 2:20; 16:13; 19:20; 20:10). The necessity of testing spirits ensures perseverance of faith that grants eternal life (1 Jn. 4:1-6; Acts 13:6). The problem is not only the simple occurrence of false prophets but much more of their ability to disguise as "angels of light" (Mt. 7:15-6; 2 Cor. 11:14). A Christian attitude shall not lead to avoidance of prophecy but rather to testing it (*dokimazein*) and to holding firmly to the good (1 Thess. 5:19-22). 1 Cor. 12:3 (Jesus is the Lord) and 1 Jn. 4: 2-3 (antignostic) serve in the early Church as means for differentiation.

The question about possibilities of spiritual discernment relates to two fundamental questions: who can discern what is true and false and what are the criteria for discernment. According to the New Testament, everything that originated from God can be known by its good fruits and vice versa (Mt. 7:15 ff.), which means, some sort of spiritual experience is necessary. The ability to discern spirits (*diakrisis pneumaton, discretio spiritum*) together with the prophecy are seen as a charisma (1 Cor. 12:10), and if used properly, they have a practical implication for the mutual construction of a Christian life individually and even more for Christian community (Hense, 2010).

The development of the concept of spiritual discernment became a part of patristic tradition in the exegetical tradition on 1 Cor. 12:10 and was widened in the monastic practice of *consider-*

atio or *examen conscientiae* (*self-examination*), (Corcoran, 1993). Many arguments were related to the debate on the origin of ideas in a believer: whether the source of prophecy, visions and dreams is God, an angel, the human spirit or an evil spirit. Modifications appeared foremost in the answer who is capable of spiritual discernment: if every monk has this capability, or only those gifted by Spirit (*charisma*), or those who gained the ability to discern as a result of an ascetic practice (Anderson, 2011, Benke, 1991).

In the long transition from *Didache*, Origen, the *Vita Antonii*, Cassian, in the West particularly in Augustine, Gregory the Great, Dionysius Carthusianus, Bernhard, Johan Gerson, Katharina von Sienna, Johan Tauler, Ignatius of Loyola, the aim of the spiritual discernment became oriented toward a general ability to differentiate and make judgement on spiritual matters (not only to discern between the spirits as in 1. Cor. 12: 10). Connected to the Aristotelian moral theory, a special virtue in this regard was the *discretio* able to avoid excesses of too much and too little. Practically it appeared as the willingness to moderation as well as to acceptance of the advice of brothers in the community (f.E. the *Rule of St. Benedict*).

In Western medieval Church the debate continued, who has the authority of discernment. The *discretio spiritum* was ascribed either to theologians or to the Church and its clerical offices or – as originally – to the monastic way of life. In other words, the criteria for discernment depend rather on intellectual or spiritual capabilities of the person.

Gregory the Great equated the prophets of the Old Testament with theologians – doctors of the Holy Church – battling infidels and heretics. This emphasis on intellectual approach to discernment between true and false is apparent in the proposal of the chancellor at the University of Paris, John Gerson. He wrote several treatises related to criteria of discerning between persons claiming a spiritual insight. His three major treatises are *De distinctione verarum revelationum a falsis* (1401–2), *De probatione spiritum* (1415) – an instrumentarium for judging on revelations and visions of Brigitta of Sweden at the council of Constance in 1415. Likewise, worth mentioning is his treatise *De revelatione doctrinarum* in 1423 (Anderson 2011, Weismayer, 1990). Gerson´s hexameter became well known: "To whom was the revelation made? What does it contain and say? Why is it said to have happened? To whom was it presented for advice? From whom and from whence did the revelation come?" (Gerson, 1987). Gradually, the spiritual discernment became the domain of authorised clerical offices in the Church from the eleventh century.

Ecclesiastical centralisation and growth of papal power by Innocent III (his letter *Cum ex iunctio*, which later became part of the *Canon Law*) and the Fourth Lateran Council made Catholic unity and doctrinal orthodoxy dependent on the properly ordained clergy and sacrament of Eucharist. From the fifteenth century, an absolute distinction between the "visionary" and the "examiner" grew gradually. The process became more rigid and qualifications about the need for experiential certitude lost importance. Wendy Love Anderson observed: "In the fifteenth and sixteenth centuries, following a judicial reading of Gerson's work, the discernment of spirits became increasingly a process of institutionally grounded examination rather than a private self-examination or an ineffable experience" (Anderson., 2011).

Monastic tradition cultivated the spiritual experience in the form of an intensive Christian faith and devoted life. Discernment of true and false could not be divided from the judgement on the results (or fruits) of a truly good and spiritual virtue: namely humility, *discretio*, patience, truth and love. Works of Cistercian abbot Bernard of Clairvaux exercised great influence in medieval and early modern time. Especially relevant are his *Sermones de diversis 23, 24, SC 32, 64* and the mystical *Sermons on the Song of Songs* (Elm, 1994). His goal was to cultivate the ability to discern the good and the truth in order to love the truth and to live according to it.

In his *SD 24*, the fundament for spiritual discernment was the Bible and the liturgy in the monastic community (Bernardus, 1994, Benke, 1991). In his treatise *De gradibus humilitatis et superbiae* the means of discernment are meditation, prayer and contemplation. Virtue of humility became a precondition for achieving the ability to discern (Bernardus, 1994). For Bernhard – as well for Ignatius four centuries later in his *The Spiritual Exercises* 4, 102, 137, 191, 201, 21 (Ignatius, 1951) – a Christian person achieves deep spiritual understanding only in conformity with Christ (*conformitas cum Christo*) particularly in his *kenosis* (the poor, humbled Jesus) on the cross. Nevertheless, Bernard was aware that even an active spiritual life does not grant a clarity in each situation. It can be difficult to discern between one's own sinfulness (*malum innatum*) and an influence of evil (*malum seminatum*) in a person (see SC 32:6, DS 23:4). The solution was to look at the spiritual fruits of a person. The life according to Christian virtues gives a hint for discernment. The criteria for discernment are practical, not primarily intellectual: "Whenever there is a salutary notion about the discipline of the body, humility of the hearth, preservation of the unity and the practice of the brotherly love or about the beginning, preserving and growing of virtues in hearth, then without doubt it is the Spirit of God speaking here. It is either He himself or He speaking through his angel" (SD 23:5, Cf. SD 24:1). In his famous differentiation of five kinds of love (SD 10:2), Bernhard advocated love as a willingness and practice of fraternal life in community (*ad faciemus bonum*). Love leads to the desire to build up the communion of believers (1 Cor. 14:1; 1 Jn. 4:6) too. For Bernardus the discernment of evil and good appears as the virtue of *discretio*, avoiding extreme positions as well as the virtue of "ordered love" (*caritas ordinatio*). Evil can be known as a contrast to good fruits: evil thoughts, desolation, quarrel, an overall negative appearance. If two or more converge together, the certainty of evil rises.

Another deeply personal and experiential way of spiritual discernment present *The Spiritual Exercises* of Ignatius of Loyola. Here Ignatius' knowledge of the Carthusian tradition, especially Dionisius and Ludolph of Saxony, of Bernhards sermons, as well as his personal experiences came to an elaborated manual of spiritual discernment (Bakker, 1970).

Ignatius was not discerning the will of God in itself but within specific experiences of "consolation" or "desolation" (Schneider, 1978). The goal of discernment is not to know the origin of spirit (evil or divine), rather its aim is practical: the recognition of the trustworthiness of the personal experience of vocation and its conformity with the will of God. It starts with putting oneself over time and space in narratives of Jesus. The process follows as a discernment of internal motives of temptations "to lack courage", "to return to former life" (SE 318-22), "of acute danger of pride" (SE 322-324a), of "false humbleness and ungrounded fear" (SE 324b-25) and "scruples" (SE 345-351). To the fourth stage of "consolation without a reason" (SE 330, 336) relates the "first stage of choice"

(SE 175-177) which is nothing else than the inner knowledge of the Lord Jesus Christ in order to love him and to follow him more (SE 104, 113, 120, 139, 147, 168), (Bakker, 1970, Schneider 1987).

The aim of discernment is less an ability to differentiate spirits and more a way to become one with the life of Christ, to follow earthly Jesus in poverty and to become his apostle in the service of preaching in the world (SE 169-188). The criterion for the progress of the exercitant became the experience of suffering with Jesus and the hope for consolation. Ignatian spirituality aimed at subordination of human activity to the service of Christ and of the Church: "[A] stable motivation to true poverty, connected with consolation and inner joy, as well as with the wish to become similar to Christ, can be accepted as a sign of divine calling" (Schneider, 1987). Spiritual peace is the definitive and decisive point in making decisions. The discernment in this approach is not given to everyone, but *The Spiritual Exercises* offer a way how to achieve it.

Theological Discernment

In the realm of theology, two developments became influential. They stemmed from Augustine's theological analysis of the problem of evil and his hamartiology – teaching about human sinfulness. An excellent analysis of Augustine's development of teaching on evil as *privatio boni* in the struggle with Manicheans offer Hermanni and Torchia (Hernani, 2010, Torchia, 1999). It is not possible to present the whole development of Augustine's teaching on sin here too (von Ort, 1989, Verschooren, 2002, O'Connell, 1969, Löhr, 2007). It shall only be noted that Augustine's thinking developed in real struggles for the goodness of creation (Manichaeism) and human partaking in it, particularly in the context of soteriology (Pelagianism). According to Ferdinand Reisinger the *Rule of St. Augustine* is full of social psychology and social responsibility as a sort "option for the weak ones" to the sick, weak and doubting. The famous and often quoted sentence appears to be a result of spiritual discernment: "With love to human being, but with hate toward the sin" (Rule of Augustine IV/10, Reisinger, 1991). The controversy with

Julian brought Augustine to the final development of his thought: "He looked afresh at the problem of evil as he had first conceived in the light of the Manichaean teachings, and he was able to set these early reflections in the larger context of sin and redemption with which he had been concerned of late" (Evans 1990).

From the vast amount of texts, the following quotation expresses the position of mature Augustine in the best way. Towards the end of his life, in 421 in his treatise *Against Two Letters of the Pelagians,* Augustine described the Catholic position by discerning theologically between good and evil theologically. Topics of Creation (good nature), Redemption (sin and grace), Sacraments and Christology form a doctrinal unity. Augustin showed the relevance of theological discernment based on distinction of sin and grace and could differentiate between three opposing theological positions: "The Pelagians do not admit that God is the Savior of a fallen mankind; the Manichees do not admit that God is the Creator of all natures; the Catholics cannot associate himself with either position, for both views diminish the Deity. The Pelagians regard the lust of the flesh as a natural good; the Manichees thing that it has been an evil thing from all eternity; the Catholics date the evil in it from the fall of Adam. The Pelagians say that even a wicked man can do good by his own free will; the Manichees say that it is from a man's free will that evil takes its beginnings; the Catholics maintain that each man is the source of his own evil (in his will) and that no-one can do good of himself. The Pelagians say that the souls is without sin, because man's nature is wholly good; the Manichees say that the soul is a particle of God, made sinful by the admixture of an evil nature; the Catholics deny that the soul is a particle of God, but they do not agree with the Pelagians that it is without sin in this life. Manichees and Pelagian agree in rejecting the notion that man's salvation depends upon the grace of Christ; they are alike in undervaluing baptism, the Manichees saying its usefulness to anyone, the Pelagian its usefulness to infants, who are without sin. They agree in doing dishonor to the flesh of Christ, the Manichees by their blasphemies concerning his birth, and the Pelagians by considering the flesh of those he redeemed equal to his own, in its freedom from original sin" (Evans, 1990).

In the sixteenth century, the reformers discovered Augustine anew and the theological discernment of sin and grace became a discernment of the "goodness of the discerner". Gillian Russell Evans formulated it pointedly: "With Luther and Calvin especially, emphasis was placed afresh on the impotence of man, the efficacy of grace. The doctrine of justification by faith alone – and faith as the gift of God – was no mechanical speculation; it was the impassioned cry of men who, like Augustine, had agonized their way to a solution and whose attempts to impose order on experience kept the Augustinian dilemma alive to the modern world" (Evans, 1990).

Augustine´s anti-Pelagian treatises have had enormous impact on the theological development of young Martin Luther. After reading a volume of Augustine's anti-Pelagian writings in Amerbach's reprint (1506) during his *Lecture on Romans* (1515–1516), Luther departed utterly from the medieval teaching about (theological) virtues as well as from Nominalist hamartiology. Interpreting Augustine´s treatise *Contra Julianum*, Luther equated concupiscence with original sin and reasoned about human passivity in the process of salvation, since from his own natural powers the sinner is unable to judge in spiritual matters properly. The sinner is not even able to feel remorse or feelings of guilt from its own powers. This position was condemned in the papal bull *Exsurge domine* in 1520 (WA 7: 103.9–111.11).

Luther went on in developing his understanding by studying Biblical terminology related to sin. The authority of Scripture, especially in his expositions of Psalms, became the instrument for theological discernment. He gradually equated the Hebrew term *hattaa* with "original sin" in the meaning of a fundamentally wrong attitude of heart. Sin as *peccatum radicale* is not a moral wrongdoing, resp. lack of virtue. It is a broken relation to God. With deepened understanding of Gospel as a promise of God that can be apprehended only through faith, Luther gained the conviction of *original sin* as unbelief in God, a distrust in Christ's promises: "[F]or nothing makes a man good except faith, or evil except unbelief" (WA 7:62.25–26, Dalfert, 2010).

The doctrine of sin belongs, according to Luther, to the "subject matter of Theology" (Batka, 2014) and – in Augustinian tradition – serves as a tool for theological discernment of good and evil: sinful and just, guilty and righteous, lost and saved. In his famous *Lecture on Psalm 51* (1532), Luther ascribed the story of the Bible containing divine and truly theological wisdom of God who is the "Justifier and Redeemer of man." God approaches people who are "guilty of sin and subject to perdition" (*homo reus et perditus et deus iustificans vel salvator*: WA 40 II:328.17-18, Bayer, 2008).

The only way for human beings to get rid of guilt of sin is to trust in God's word promising salvation in Jesus Christ – which happens out of pure grace of God.

Nevertheless, this is not the end of discernment. In Luther´s theology, the solution ends in the state of simultaneous sinfulness and justice. The paradoxical – and often rejected – formulation of *simul iustus et peccator* (Wicks, 1992, Körtner, 2018) deepens the theological discernment between sin and faith. It opens a way for continuous inner reflection and judgement on every kind of human act, thought, emotion and motivation. Luther's goal was to lead every Christian to daily *metanoia* – change of thinking, affections, form of (Christian) life: "The importance of the doctrine is clear enough. But the understanding of it requires attention to the consequences Luther drew from the *simul* for daily life and prayer. In the lived religiosity of the believer, the *simul* indicates the necessary primacy of a penitential spirituality in daily practice and prayer" (Wicks, 1992). In this way the dialectics of *simul iustus et peccator* expresses the very human predicament, opening way for deep insight of the human weakness in discerning between good and evil, and for the profound human dignity in the sight of the worth of the grace and faith.

Already in his early reformational treatises *On Christian Liberty* (1520) and *To the Christian Nobility of the German Nation* (1520), Luther argued for the priesthood of all believers, or the priesthood of all the baptised. For Luther – in contrast to Bernhard – the moral (virtuous) components are less important. The ideal is not only the monastic life but also baptism and life in faith (Lohse, 1994). For him, fruits of spiritual excellence or the certainty of choice are not the means for discerning. Much more than that, it is the Word of God in its twofold form as word of judgement and of promise. Every Christian living in faith is able to discern between human and divine by means of distinguishing Law and Gospel. The Gospel deals with Christ (*was Christum treibet*, WA DB 7: 384.29-32). This means enables every Christian to interpret the Scripture. The ideal is that every Christian – male and female – has the authority to discern between good and evil. It would be a mistake to interpret Luther's position as antirationalistic. At the Diet at Worms (1521) Luther refused to revoke his teaching unless his conscience became convinced by the Scripture (*testimoniis scritpurarum*) or rational reasons (*ratione evidente*), (WA 7: 838.4-8).

The theological ability to discern is amplified in Luther's so-called "theology of the cross". In his *Heidelberg Disputation* (1518), Luther formulated a paradoxical statement (Thesis 21): "A theology of glory calls evil good and good evil. A theology of the cross calls the thing what it actually is" (Blaumeister, 1995, Forde, 1997, McGrath, 2011). Theological discernment can see – paradoxically – God hidden under weakness, suffering, or the cross (WA 5: 418.34-419.3). A theologian of the cross does not equate the appearance of "glory" with goodness, and the appearance of the "cross" with evil (Cf. 1 Cor. 1: 21). Rather, the seemingly evil event (crucifixion) can be seen as an instrument of divine action (salvation).

God uses the good creation in a good way and counterbalances human perverted use of good creation for evil goals. Again, the theological discernment touches the theological judgement on human predicament as *homo incurvatus in seipso* (equal to original sin). Those who see the "good" in the cross (through faith) lose their blindness (sin) and gain understanding of the real conditions (sin and faith).

Luther's immersion in Biblical language gave him the opportunity to express the wide spectrum of broken human relationships to God, to the created world, and to oneself. Besides classical passages like Gen. 2-3; Gen. 8; Ps. 32 and 51; Mt. 7:17-19; 12:33 and Rom. 1-7, important Biblical texts became Mt. 7:16-20 and 12:33, Lk. 6:43-45. The parable about a good tree bearing good fruits and bad tree bearing bad fruits – important for spiritual discernment by Bernhard – offered a variety of motives illuminating the human condition in a theological way. Naturally, people do not understand themselves as sinners. But sin makes them to suffer under its enslaving destructive power. Sin makes people powerless. By oppressing others, a sinner finally becomes a sinner being oppressed. A person without love dies by the lack of love. It is a struggle within one's self in a state of being delivered to evil. This fundamental sinful predicament should not be taken too narrowly as talking only about human moral capabilities. For Luther, the denial of sin leads not only to rejection of God's justification but also to blind self-righteousness and disability to judge good and evil properly. Theological discernment of good and evil is helping the sinner to understand him- or herself.

AI and Discernment

In October 2017, Saudi Arabia moved to a major precedent step in technological development. It awarded state citizenship to the social robot Sophia, developed by the company Hanson Robotics. This humanoid "hybrid human-AI intelligence" robot is operating in a fully AI autonomous mode, and other times its AI is intermingled with human-generated words. Her family of human developers (engineers, artists, scientists) wants to craft and guide her conversations, behaviours and her mind. The sentience is both an AI research project and a kind of living science fiction, driven by principles of character design and storytelling, cognitive psychology, philosophy and ethics, used to conceptually explore her life's purpose in this time of accelerating change. The team at the Hanson Robotics Corporation consists of widely diverse expert AI scientists, philosophers, artists, writers, and psychologists from diverse cultures, ethnicities and gender orientations, working together towards the ideal of humanising AI "for the greater good" (Hanson Robotics, 2019).

We can ask, what is the notion of good implied here? The benefit to human beings or all creatures on this planet? Are we dealing here with the perception of good by human constructors of Sophia? It is necessary that AI should be able to discern between good and evil.

Perhaps, it might be easier in the realm of morality or jurisprudence. In April 2019, an independent Commission of Experts at the level of European Commission presented ethical guidelines for "Building trust in human-centric artificial intelligence" (EK, 2019) which include ethical design and moral standards. It is important to prevent the "encapsulation of physical gender stereotypes in technology and robotics" since the "technology captures and reproduces controlling and restrictive conceptions of gender and race which are then repetitively reinforced" in working machines (Kišoňová, 2021). The use of AI in the realm of jurisprudence seems not to be fiction. From the early 1980s considerable work was done on developing legal expert systems. AI technologies can enable lawyers to serve more clients more effectively at lower cost, increase access to justice by serving the legal needs of the poor, and focus time and expertise on work that requires the uniquely human skills of empathy. AI is already present in areas of legal research; contract analysis; case prediction; learning how judges think, write and rule; and comparing the expertise, performance and success of law firms, (compare companies like *Fastcase, Ravel Law, DataLex* project). Some advocate that AI take over the role of judges in the societies.

Can AI become an independent agent, able to discern between good and evil in full complexity including the discernment between good and evil in the spiritual or theological way?

When we speak about the spiritual discernment of good and evil as differentiation between true and false, it might be possible for AI to make a rational discernment. AI could make decisions that resemble virtues, humility even acts of altruism – "love" as "good fruits". AI could be able to make discernment in spiritual matters, for example, about its conformity with the will of God or life of Christ, however, without the inner experience of vocation or joy or spiritual peace. AI can write prayers or sermons, AI is able to speak a prayer, but until it is a subject, AI will not be able to pray in spirit.

As to the theological discernment, AI would actually resemble a person in the state of original sin. Without divine intervention, it cannot develop remorse, feeling of guilt or a relationship to divinity, called "faith". Theologically, evil is conjoined with sin as a negative result of lack of trust in God. The discernment between good and evil shapes the dynamics of human life in a very broad sense. Theological discernment is able to speak about good and evil beyond moral categories. Sin as unbelief and liability to self-destructive structures, the aspect of personal responsibility for evil and the aspect of the tragic deliverance of this evil describe a complex state of human predicament – from which the human production of AI is not immune.

In realms of relationships to God and other people, and in awareness of the quality, complexity and ambiguity of relationships, human beings gain a greater complexity than any other creature (AI included). AI can become an independent agent but cannot discern between good and evil in full complexity of spiritual and theological realms.

We can conclude with a conviction that every thought about good and evil serves as the protection of human value and human dignity. It could be that the theological discernment remains the last resort for the humaneness in its proper sense.

References

Anderson, W. L. (2011). *The Discernment of Spirits. Assessing Visions and Visionaries in the Late Middle Ages*. Tübingen: Mohr Siebeck.

Bakker, L. (1970). *Freiheit und Erfahrung. Redaktionsgeschichtliche Untersuchungen über die Unterscheidung der Geister bei Ignatius von Loyola*. Würzburg: Echter.

Batka, L. (2014). Luther's Teaching on Sin and Evil. In: Kolb, R., Dingel, I., Batka, L. (eds.). *The Oxford Handbook of Martin Luther. Theology* Oxford: Oxford University Press, 233–253.

Bayer, O. (2008). *Martin Luther's Theology. A Contemporary Interpretation*, translated by Thomas Trapp. Grand Rapids: Eerdmans Publishing Company.

Benke, Ch. (1991). *Unterscheidung der Geister bei Bernhard von Clairvaux*. Würzburg: Echter.

Bernardus. (1994). *Sämtliche Werke: lateinisch/deutsch*. In Gerhard Winkler (ed.). Würzburg: Tyrolia.

Blaumeiser H. (1995). *Martin Luthers Kreuzestheologie*. Paderborn: Bonifatius.

Corcoran, D. (1993). Geistliche Führung. In McGinn, B., Meyendorff, J., Lexlercq, J. (eds.). *Geschichte der christlichen Spiritualität*. Vol. 1, Würzburg: Echter, 437–443.

Dalfert, I. (2010). *Malum. Theologische Hermeneutik des Bösen*. Tübingen: Mohr Siebeck.

Elm, K. (1994), *Bernhard von Clairvaux. Rezeption und Wirkung im Mittelalter und in der Neuzeit*. Wiesbaden: Harrasowitz.

Evan, G. R. (1990). *Augustine on Evil*. Cambridge: Cambridge University Press.

Forde G. (1997). *On Being a Theologian of the Cross: Reflection on Luther's Heidelberg Disputation, 1518*. Grand Rapids: Eerdmans Publishing Company.

Gerson, J. (1987). *Opera Omnia*. Ed. Louis Ellies Du Pin. Hildesheim: Georg Olms Verlag.

Hense, E. (2010). *Frühchristliche Profilierung der Spiritualität. Unterscheidung der Geister in Ausgewählten chriften.* Berlin: LIT.

Hermanni, F. (2010). Augustinus über Gott, das Gutsein des Seienden und die Nichtigkeit des Bösen. In Berges, B., Hermanni, F. (eds.). *Augustinus. De natura Boni – Die Natur des Guten.* Paderborn: Schöningh, 47-68.

Kišoňová, R. (2021). The Identity and Gender Problem in the Context of AI. *Human Rights: From Reality to the Virtual World.* Józefow: Alcide De Gasperi University of Euroregional Economy in Józefów, 43-55.

Körtner, U. (2018). *Luthers Provokation für die Gegenwart. Christsein, Bibel, Politik.* Leipzig: Evangelische Verlagsanstalt.

Lohse, B. (1994). Luther und Bernhard von Clairvaux. In Elm, K. (ed.). *Bernhard von Clairvaux. Rezeption und Wirkung im Mittelalter und in der Neuzeit.* Wiesbaden: Harrassowitz, 271-301.

Lohse, B. (1999). *Martin Luther's Theology. Its Historical and Systematic Development.* Edinburg: T&T Clark.

Lohr, W. (2007). Sündenlehre. In Drecoll, V., H. (ed.). *Augustin Handbuch.* Tübingen: Mohr Siebeck, 498-506.

Luther, M. (1883). *D. Martin Luthers Werke.* Kritische Gesamtausgabe. Weimar: Hermann Böhlau. (WA)

McGrath A. (2011). *Luther's Theology of the Cross: Martin Luther's Theological Breakthrough.* 2nd. ed. West Sussex: Wiley-Blackwell.

O'Connell, R. (1969). *St Augustine's Early Theory of Man, A.D. 386-391.* Cambridge Massachusetts: Harvard University Press.

Reisinger, F. (1991). Augustinus. In Weismayer, J. (ed.). *Mönchsväter und Ordensgründer. Männer und Frauen in der Nachfolge. Jesu,* Würzburg: Echter, 49-64.

Sancti Aureli Augustini. (1956). *Corpus Scriptorum Ecclesiasticorum Latinorum* (CSEL). Wien: Akademie der Wissenschaften.

Schneider, M. (1987). *"Unterscheidung der Geister". Die ignatianischen Exertitien in der Deutung von* In E. Przywara, K. Rahner und G. Fessard. 2. (eds). Innsbruck: Tyrolia.

St. Ignatius. (1951). *The Spiritual Exercises of St. Ignatius: Based on Studies in the Language of the Autograph.* Translated by Louis J. Puhl. Chicago: Loyola University Press.

Torchia, N., J. (1999). *"Creatio ex nihilo" and the Theology of St. Augustine. The anti-Manichaean polemic and beyond.* New York: Peter Lang.

Verschooren M. (2002). The appearance of the concept concupiscentia in Augustine's early antimanichaean writings (388-391). *Augustiniana,* 52, 199-240.

von Oort, J. (1989). *Augustine on Sexual Concupiscence and Original Sin: Papers presented to the Tenth International Conference on Patristic Studies 4,* ed. by E. Livingstone Leuven: Peeters Press, 328-386.

Weismayer, J. (1983). *Leben in Fülle. Zur Geschichte und Theologie christlicher Spiritualität.* Innsbruck: Tyrolia.

Weismayer, J. (1990). Ein Blick in einen fernen Spiegel. Spätmittelalterliche Traktate über die Unterscheidung der Geister. In Imhof, P. (ed.). *Gottes Nähe. Religiöse Erfahrung in Mystik und Offenbarung.* Würzburg: Echter, 110–126.

Wicks, J. (1992). *Luther's Reform. Studies on Conversion and the Church.* Mainz: Verlag Philipp von Zabern.

Zouhar, M. (2016). Konceptuálna analýza v analytickej filozofii. *Filozofia,* 5, 410–424.

The Relevance of Legal Intuitionism and Selected Moral Emotions in Legal Thinking and Decision-Making Processes

Olexij M. Meteňkanyč

Abstract. Philosophy, including the legal one, has long been dominated by rationalist models of thinking. Rationalist approaches in philosophy stress the power of a priori reason to grasp substantial truths about the world. This is reflected in legal thinking in many ways, where we place considerable emphasis on the legal system and its application being logical, systematic, objective and certain. Some truths, dogmas and even, dare we say, myths about the functioning of law, traditionally adopted and passed on from one generation of lawyers to the next, are still being taught at law schools. However, nowadays we understand that the rationalistic view of human knowledge in law described above is not the only one, and it is possible to oppose it with alternative views on the functioning of the human mind and, consequently, of legal thinking as well. In this chapter, we will attempt to show that the decisions of the judges and other bodies applying the law are not merely an act of rationality based on the rule that judges logically and mechanically apply legal norms to the facts of the cases that appear on their table (with the help of a legal syllogism). In our opinion, intuition and experienced (moral) emotions significantly influence the decision-making process, and it is necessary to take this into consideration. With this chapter, we primarily want to contribute to a kind of demystification of legal thinking, with a particular emphasis on decision-making processes in law. In particular, we want to focus on the requirement of excessive rationality in the application of law that we often read in legal theory textbooks. Therefore, on a conceptual level, this chapter combines two interdisciplinary approaches: on the one hand, the selected aspects of legal intuitionism as presented by American legal realists, especially as formulated by those from the "psychological" branch of this movement (J. Hutcheson, J. Frank); on the other hand, we have the recent concept of moral emotions (especially as presented by Jonathan Haidt), which significantly (on an empirical basis) analyses intuitions and emotions in the formation (not only) of moral judgements in human decision-making processes. We believe that selected aspects of both interdisciplinary approaches will be useful in revealing the place of intuition and emotion in the decision-making processes of judges and other bodies applying the law, as well as in legal thinking as such. In conclusion, we will present a number of practical implications that can be drawn from the examined approaches, which we believe can be an enrichment for legal scholars and the legal community (especially for our Slovak, and even Central European, region).

[W]hen the case is difficult or involved, and turns upon a hairsbreadth of law or of fact (...) I, after canvassing all the available material at my command, and duly cogitating upon it, give my imagination play, and brooding over the cause, wait for the feeling, the hunch – that intuitive flash of understanding which makes the jump-spark connection between question and decision, and at the point where the path is darkest for the judicial feet, sheds its light along the way.

(Hutcheson, 1929, 278)

Introduction

Legal thinking is a phenomenon that is difficult to grasp. Like law itself, it has many layers, twists and dead ends, but it has been at the centre of our interest for centuries. However, it is quite interesting to see that some issues in a particular area are given significant attention (perhaps too much in some instances) by legal theorists and philosophers, while other topics are neglected. In this chapter, we would like to focus on one of those seldom explored topics (at least in our Slovak, or even Central European region). We believe that the concept of moral emotions and intuitions in legal thinking (especially in the field of application of the law) is not given sufficient attention, despite the development of knowledge in cognitive sciences (and moral psychology in particular).

Why is this so? Generally, we believe that emotions and intuitions are perceived in the legal environment as irrational, unsystematic and above all subjective elements. However, these adjectives do not correspond to the traditional presentation of how law and legal thinking are supposed to work. In our environment, there is a significant emphasis on rationality. Rationalism has many meanings in philosophy (see, e.g., Dea et al., 2018; Markie, Folescu, 2021), but in this chapter, we rely on its rather moderate concept: it is the view that reason is the main source of valid knowledge in both ethics and law. Philosophy, including the legal one, has long been dominated by rationalist models of thinking. Rationalist approaches in philosophy stress *"the power of a priori reason to grasp substantial truths about the world"* (Williams, 1967, 69). This is reflected in legal thinking in many ways, where we place considerable emphasis on the legal system and its application being logical, systematic, objective and certain. Some truths, dogmas, and even, dare we say, myths about the functioning of law, traditionally adopted and passed on from one generation of lawyers to the next, are still being taught at law schools. This is especially true in our Central European region, where many of the ideas of the dogmatists or the normativists, both legal-philosophical trends that place considerable emphasis on a certain type of rationality, still resonate to this day (on dogmatism in law see, e.g., Harvánek et al., 2008, 36–37; Turčan, 2020, 13–19; on legal normativism in turn see Weyr, 2015; Kelsen, 2018; Colotka, Káčer and Berdisová, 2016, 13–27).

However, nowadays we understand that the rationalistic view of human knowledge in law described above is not the only one, and it is possible to oppose it with alternative views on the functioning of the human mind and, consequently, of legal thinking as well. One of them relies on intuitions and moral emotions, and the view given is exemplified by the opening statement of this chapter, which was presented in the last century by Justice Joseph Hutcheson in his essay (1929) on intuitive decision-making in law. Indeed, Hutcheson presents a diametrically different way of deciding cases by courts and other bodies applying the law, which differs significantly from the rationalist version. He states himself that *"I speak now of the judgment or decision, the solution itself, as opposed to the apologia for that decision (...) I speak of the judgment pronounced, as opposed to the rationalization by the judge on that pronouncement"* (Hutcheson, 1929, 279). Thus, in his understanding, there is a distinction between two processes that are often perceived (especially by rationalists) as one. A decision and the reasoning behind it (justification) are not a single act of rational thought. The decision often comes first, and on the basis of educated intuition, the reasoning then follows with a reflection on existing standards of what a good judge's decision should look like.

From our point of view, this is a significant moment that is essential to the so-called intuitionist theory of decision-making in law. Through its lens, the importance of legal intuition in the decision-making process of judges is pointed out, which can be influenced by several subjective factors (legal, political or moral attitudes of the judge, as well as by his or her upbringing, education, the environment in which he or she grew up, etc.), which refutes the myth existing in legal thinking that the application of law is, or should always be, a logical, certain and purely rational process, which is supposed to have the following structure:
1. analysis of the facts of the case;
2. finding and interpretation of the applicable law;
3. conclusion reached by applying the legal syllogism (Prümm, 2016; Gábriš, 2020).

In this chapter, we will attempt to show that the decision of the court and other bodies applying the law is not merely an act of rationality based on the rule that judges logically and mechanically apply legal norms to the facts of the cases that appear on their table (with the help of a legal syllogism). In their decision-making, they are not guided only by the letter of the law or, from their point of view by the correct interpretation of the law; on the contrary, in our opinion, intuition and experienced (moral) emotions significantly influence the decision-making process and it is necessary to take this into consideration.

On a conceptual level, this chapter combines two interdisciplinary approaches: on the one hand, the selected aspects of legal intuitionism as presented by American legal realist representatives, especially as formulated by those from the "psychological" branch of this movement, i.e., not only the aforementioned Joseph Hutcheson but also Jerome Frank, both well-known jurists and judges of the first half of the twentieth century; on the other hand, we have the recent concept of moral emotions (especially as presented by Jonathan Haidt), which significantly (on an empirical basis) analyses intuitions and emotions in the formation (not only) of moral judgements in human decision-making processes. We believe that selected aspects of both interdisciplinary approaches will be useful in revealing the place of intuition and emotion in the decision-making processes of courts and other bodies applying the law. We shall commence with the first mentioned, namely legal intuitionism as presented by American legal realists.

Legal Intuitionism According to Selected American Legal Realists (A Historical Theoretical Perspective)[6]

The question of the existence and relevance of legal intuition in judicial decision-making was given quite considerable attention in the first half of the twentieth century in the United States, in part thanks to a select group of American legal realists belong-

---

6 The following part of the text was partly published in Slovak language in the past in Meteňkanyč (2018, 207–220).

ing to the so-called psychological branch.[7] Legal intuition was already pointed out by Benjamin N. Cardozo, who in several places in his work *The Nature of the Judicial Process* emphasises an idea that a judge is not a mechanical machine, who follows only the dictates of logic and law, and makes his decisions considering several other factors, including legal intuition (Cardozo, 2011, 59, 93–95, 126–130). A judge's thinking is mainly influenced by psychological forces that are largely subconscious. *"Deep below consciousness are other forces, the likes and the dislikes, the predilections and the prejudices, the complex of instincts and emotions and habits and convictions, which make the man, whether he be litigant or judge."* (Cardozo, 2011, 127). The aim of Cardozo's lectures, summarised in the above-mentioned work, is to give a practical answer to the question "What is it that I do when I decide a case?", emphasising an important "banality" that is often forgotten, namely that a judge is a human being like any other. In this regard, Cardozo sums up: *"We may try to see things as objectively as we please. Nonetheless, we can never see them with any eyes except our own"* (Cardozo, 2011, 31). To put it another way: the judge is also unable to cut him or herself completely off from his or her emotions and his or her own nature when making decisions. It is not possible for him/her to transcend his/her ego and see things as they really are (Sobek, 2011, 323).

Currently, however, Judge Hutcheson is the most frequently cited as a representative of legal intuitionism among the proponents of American legal realism. It should be noted that Hutcheson was initially an example of a judge of legal classicism (the orthodox

---

[7] Nevertheless, it is rather a marginal branch of American legal realism, although it is often presented as a typical representation of this movement. Even legal realists themselves have criticised this understanding of legal intuitionism. As an example, Felix Cohen criticised intuitionists for focusing too much on the individual psychological factors of judicial decision-making ("on the person of the judge") while overlooking the fact that judicial decision-making is co-determined by a variety of objective social factors. According to Cohen, judges are more influenced by facts such as the ability of judges to communicate with each other, share a common environment, adhere to the traditions of their predecessors and mentors, work with the same sources of law, have the same education, or publish their experiences in the same professional journals and present them at the same conferences. Cohen sees the individuality and subjectivism of the judge as secondary to the aforementioned objective social determinants (Cohen, 1935, 843 et seq., alternatively see also Sobek, 2011, 327).

direction), i.e., he considered legal thinking to be logical and rational thinking, with precisely defined legal categories and institutions, in which the judge's personal feelings and emotions, as well as other extra-legal factors, do not belong, and are something unscientific and unprofessional if taken into account by the judge.[8] It was only with increasing years of experience on the bench that he began to realise that the application of law is not only based on logical legal thinking but that judges are also guided by their "hunches" and intuition when deciding cases. This eventually led him to accept the well-known conclusion that *"the judge really decides by feeling, and not by judgment; by 'hunching' and not by ratiocination"* (Hutcheson, 1929, 285).[9]

But would it not be startling if Hutcheson, having been convinced that the process of applying the law is a logical and rational process, suddenly became an advocate of some random, intuitive, case-by-case decision-making process? Not exactly, since he was already reflecting the complexity of judicial intuition at the time. In his perception, it too has its criteria, it is not random, but it takes shape with the judge's increasing experience in deciding cases. He points out that the intuitive decision-making process itself should have two phases. First, the judge must familiarise himself with all the existing and available information (factual and legal) and only then let his intuition (sense of appropriateness) guide him or let his "legal sense" nudge him towards the right conclusion of the case. According to Hutcheson, an intuitive feeling for the right decision in a particular matter is a vital, motivating impulse for decision (Hutcheson, 1929, 274 and 285). Metaphorically speaking: intuition is the skipping of the spark between the question being asked and the decision being made, and it can be especially seen in difficult cases that are tangled. In these cases, after carefully examining

---

8 *"I had been trained to regard the law as a system of rules and precedents, of categories and concepts, and the judge had been spoken of as an administrator, austere, remote, 'his intellect a cold logic engine', who, in that rarified atmosphere in which he lived coldly and logically determined the relation of the facts of a particular case to some of these precedents (...)"* (Hutcheson, 1929, 274).

9 At the same time, however, Hutcheson was aware that judges do not usually speak of intuition as a decisive factor in their decision-making, since *"(judge) would be stoned in the street if it were known that, after listening with full consciousness to all the evidence, and following as carefully as he could all the arguments, he waited until he 'felt' one way or the other"* (Hutcheson, 1929, 282).

the materials and reviewing the legal aspects of the case, the judge should *"wait for the feeling, the hunch – that intuitive flash of understanding which makes the jump-spark connection between question and decision, and at the point where the path is darkest for the judicial feet, sheds its light along the way"* (Hutcheson, 1929, 278).

Nevertheless, it should be emphasised that Hutcheson does not dismiss the importance of legal doctrine; on the contrary, it has its importance, especially in the gaining of experience of young judges. In practice, the role of doctrine is not to exclude the personal convictions and inclinations of judges but to provide them with the means of testing their intuitions about the best judgement against the concerns and results reached by other judges in other cases (Powell, 2009, 1726). However, the intuition itself, after gaining enough experience, begins to prevail. At the same time, he perceived the fact that intuitions are more subtle in distinguishing legally significant distinctions than any precisely formulated premise of an argument (Sobek, 2010, 176). The judge first makes a decision that is intuitively acceptable to him or her and only then tries to formally justify his or her decision in legal categories (i.e., on the basis of certain legal rules, analogies, concepts and institutes). Thus, to sum up, according to Hutcheson, the judge's decision itself is intuitive, but once it is pronounced, it must conform to the prevailing legal thinking, emphasising logic and rationality. While a judge may *de facto* reach his or her decision in a case intuitively, he or she is forced to subsequently compose his or her official reasoning for the decision, which is essentially a *post-hoc* rationalisation of the intuitive opinion. Here, it can be seen that these are two independent actions, each playing a significant role in the judge's decision-making processes. Yet Hutcheson concludes his essay by pointing to those judges who can intuitively predict the best decision in a given legal case and, in the process, write the "dullest" justifications for their opinions (Hutcheson, 1929, 287). This is something to be avoided.

Interestingly, Hutcheson did not believe that judicial intuitions could be scientifically explained (Hutcheson, 1929, 287–288; Modak-Truran, 2001, 87).[10] In this, he differed substantially from another

---

10 Surprisingly, a few decades later the above, based on Hutcheson's ideas, was carried out by the contemporary renowned moral psychologist Jonathan Haidt

important representative of legal intuitionism, namely Jerome Frank.

Jerome Frank has been viewed rather controversially by many past and present-day authors (Paul, 1959, 129–143; Volkomer, 1970, 207–221; Gábriš, 2020, 47–49 and 52–54). Stępień aptly points out that Frank's legal thoughts have not been utilised for the study of court opinions to date is most likely because of the superficial readings of his ideas as an attempt to incorporate aspects of Freud's psychological theories into the legal sphere (2019, 328). Frank, though, is in many ways still inspiring to us to this day (see, e.g., Meteňkanyč, 2020), and there is no need to emphasise only his "radical opinions" which, however, may not sound so extreme upon closer examination. Let us stick, however, with his perception of legal intuitionism, where the judge's subjectivity and intuitions are fundamental to the decision-making process. These inevitably accompany the decision-making actions of judges, with uncertainty in the adjudication of cases as a consequence (which Frank does not perceive as a negative).

Frank's understanding of legal intuition shared a number of overlaps with Hutcheson's view described above. In particular, it is worth mentioning the *post hoc* justification of decisions, i.e., judges first let their intuitions guide them and only then try to justify their decision in legal categories. Frank, however, in characterising intuition goes further than Hutcheson. Intuition is made up of several elements, but it is mainly shaped by the legal, political, economic and moral attitudes of the judge, his/her upbringing, education, character and so on.[11] Basically, we are talking about the whole personality of the judge. However, he also diverges from Hutcheson in the process of finding an intuitive decision. According to Frank, the judge does not inquire into the legal standards and facts of the case and then make an intuitive assessment of

---

in his paper *Moral Psychology and the Law: How Intuitions Drive Reasoning, Judgment, and the Search for Evidence* (2013).

11 According to Frank, a judge's past and personal experience also play a big role in the decision-making process. "*His own past may have created plus or minus reactions to women, or blonde women, or men with beards, or Southerners, or Italians, or Englishmen, or plumbers, or ministers, or college graduates, or Democrats. A certain twang or cough or gesture may start up memories painful or pleasant in the main*" (Frank, 1949, 106).

everything in summary. The whole process works exactly the other way around. First, the judge makes an intuitive decision and only then looks for legal and factual reasons supporting his decision (Frank, 1949, 100). In some cases, a decision reached in the above manner is not sufficiently supported by the law or the facts of the case, and the judge must reconsider his decision, i.e., if, in Frank's words, he/she is not *"arbitrary or mad"* (Frank, 1949, 100).

Simultaneously, with this position, Frank points out that judicial decisions are reached in a different way than traditionalist legal theories of decision-making (based on logical/mechanical reasoning) assume. For a better understanding of his position, we can mention Frank's article *Say It with Music*, which is a rather strange-sounding title for a scholarly paper presented by a judge and a legal practitioner. In this paper, Frank presents partly his reflection on the nature of judges' decision-making processes. To better capture how judicial decisions are made, Frank refers to the concept of *gestalt* (Frank, 1948, 928), the "wholeness" of both the judge's experiences during the case that influence the decision-making process and the judge's response towards the case. What is crucial here is that the *gestalt*-like characteristic of judging, *"the non-analytical character of the decisional process"* (Frank, 1948, 928), influences other aspects of the decision-making process and the final version of judicial opinions. One of the consequences is the existence of inherent and unavoidable problems with the articulation of decisional premises ("real reasons" leading to the decision). Frank points out that judge *"experiences a Gestalt; that is why he has difficulty in reporting his experience analytically"* (Frank, 1948, 929). Why, however, according to Frank, is this happening? Why does the *gestalt*-like character of judging lead to serious troubles with articulating decisional premises?

In simple terms, the idiosyncratic and *gestalt*-like experiences of a judge cannot be reduced to simple propositional theses without losing something important. This is the major reason why decision-makers have a rudiment of difficulty in articulating these experiences, even for themselves. Frank operates with the notion of *"inexpressibility of the decisional process"* (Frank, 1948, 929) to highlight the problems with formulating what really happens when passing a decision in a "logical, lingual form" (Stępień, 2019, 320). Judges' responses are partly based on and bounded by their "wordless knowledge", i.e., there is always a part of the decisional premises that is "unspeakable" and so cannot be written down. In this regard, Frank relies on Professor Langer's findings, in particular citing her thoughts that *"everybody knows that language is a very poor medium for expressing our emotional nature. It merely names certain vaguely and crudely conceived states, but fails miserably in any attempt to convey the ever-moving patterns, the ambivalences and intricacies of inner experience, the interplay of feelings with thoughts and impressions, memories and echoes of memories (...) all turned into emotional stuff."* Much that we call *"intuitive knowledge"* is *"itself perfectly rational, but not to be conceived through language"* (Langer, 1948, 80–82). The above is true in relation to particular feelings, emotions and hunches but relates most especially to the holistic-type experiences, which consist of many smaller elements, some of which are impossible to put labels on (Stępień, 2019, 320). Since the judge is not engaged in a wholly logical enterprise, *"the effort to squeeze his 'hunch', his wordless rationality, into a logical verbal form must distort it, deform it"* (Frank, 1948, 932–933).

And at this point it is appropriate to explain the title of the paper itself, *"Say It with Music"*. In this respect, Frank compares the process of adjudicating court cases to the artistic process, noting a number of similarities. In both processes, the ability to discover, the necessity of creativity of the creator and, of course, the strong influence of personality on the processes of creation themselves play a crucial role. Therefore, it is possible to interpret Frank´s essay as a straightforward plea to judges to acknowledge the inherent weakness of words. Not everything can be put into words. There is a song of the same name by Irving Berlin that was popular in those days. The main idea of this song is to suggest that love is a complex and internal emotion that cannot be simply described with words.[12] It is on the basis of this analogy that Frank emphasises the importance of the nonverbal elements present during judges' decision-making processes. Music (also in the

---

12 Compare with the later 1958 version, but still an authentic one: Pat Boone – Topic: Say It With Music. In: *Youtube*, published on Jul 31, 2018. Available at: https://www.youtube.com/watch?v=zVL_87h4cSw&ab_channel=PatBoone-Topic (accessed on 25.12.2022).

sense of Berlin's song) helps us better express our feelings. And yet we are not able to give a complete representation of our precise experience of love; something always seems to have been left out, namely those elements which are the most vivid, the most essential, the most subtle, the most touching and the most transforming. Frank suggests that a judge faces similar existential problems in articulating the reasons for a decision already made (subject, of course, to certain distinctions, and the comparison in question applies only to a certain extent). His suggestion is: although there is a significant quantitative difference between a lover and a judge, judges should follow the example of the lover from Berlin's song and try to "*say it with music*" (Frank, 1948, 921–922 and 931–934).

In his descriptions of judicial decision-making, Frank often views the case not only through the lens of the judge but also from the perspective of the attorney. The true essence of advocacy mastery lies in correctly blending the arguments of a dispute. In other words, the baseline for an advocate is the need to know the reasoning and arguments that suits his/her client and appeals to/convinces the judge. For this reason, the attorney must craft an appropriate combination of factual and normative premises that will be as impressive and acceptable to the judge as possible, and it would be advisable for the attorney to include incentives to play the judge's imagination to the advantage of the attorney so that the judge's legal intuition will be directed in favour of the attorney's argumentation (Frank, 1949, 101–104). The aforementioned procedure is mainly justified by the fact that a judge is also just a human being and, according to Frank, people in their normal thought processes do not use syllogistic reasoning in their decision-making procedures, but on the contrary, they use their intuitive decision-making processes based on the subjectivism of the human being, including the judge. Therefore, it is reasonable to assume that a judge will not use artificial forms of inference simply because he has put on a judicial ermine. Judicial judgements, like other judgements, doubtless, in most cases, are worked out backward from conclusions tentatively formulated (Frank, 1949, 101). Also for this reason, according to Frank, the study of law should pay special attention to the importance of human nature in legal practice (hence human psychology should also be taught in law schools, with a focus on judicial decision-making).

Let us compare the above-mentioned views with one of the recent positions on this issue, namely, the opinions of Hutcheson and Frank[13] are criticised by the contemporary eminent theorist and judge Richard Posner in his work *How Judges Think*.[14] At the same time, however, we should add that in many ways he relates to them. First of all, it should be noted that Posner distinguishes multiple factors that are capable of influencing judicial decision-making: in addition to intuition, these are mainly emotions,[15] political feeling[16] or the ability to make a "good decision".[17] However, the most fundamental is intuition, which *"plays a major role in the judicial process, as in most decision-making processes"* (Posner, 2008, 107). Simultaneously, Posner points out that a distinction should be made between "hunch" and intuition. On the one hand, the hunch should only represent a kind of a judge's guess on the case (a shot in the dark); on the other hand, the intuition should stem from the judge's education and professional experience, respectively, and on the basis of the said experience and education, the judge should have the ability to interpret the law (Sobek, 2010, 176). *"The faculty of intuition that enables a judge, a businessman, or an army commander to make a quick judgment without a conscious weighting and comparison of the pros and cons of the possible courses of action is best understood as a capability for reaching down into a subconscious repository of knowledge acquired from one's education*

---

13 For example, Posner's criticism of Frank's suggestion that judges should undergo psychoanalytic tests and psychotherapy in order to "train" their intuition, which was not only criticised by Posner but was also not favourably received by the judges themselves.

14 For further details see specifically chapter 4 of the book entitled "The Mind of the Legislating Judge" (Posner, 2008, 93–121).

15 Particularly in relation to litigants' persons or witnesses, where *"the judge is likely to set some emotional reactions to one side, such as a personal liking for a litigant or his lawyer"*. In this respect, it is also necessary to regard the judge as just an ordinary person who has an emotional side, which he or she is unable to set aside, even though he or she is required to do so in terms of the objectivity of the case (Posner, 2008, 106).

16 That is, whether the judge is a conservative or a liberal, etc. (Posner, 2008, 100–107).

17 Posner refers to this faculty as akin to intuition, which can be characterised as *"an elusive faculty best understood as a compound of empathy, modesty, maturity, a sense of proportion, balance, a recognition of human limitations, sanity, prudence, a sense of reality, and common sense"* (Posner, 2008, 117).

and particularly one's experiences (as in 'practice to the point of automaticity')" (Posner, 2008, 107).

Posner also highlights the economic necessity of intuitive decision-making by judges in some cases. The alternative, in addition to the judge's intuitive decision, is to apply explicit, step-by-step rational reasoning based on smaller bits of information, while *"the costs of consciously processing the information may be so high that intuition will enable a more accurate as well as a speedier decision than analytical reasoning would"* (Posner, 2008, 108). This is particularly true in cases which are similar in content and which have been decided by the judge in the past. With the experience gained, a judge can decide a case much more quickly on the basis of his or her acquired skills than a "novice" judge.[18]

It is obvious that Posner attaches great importance to intuitions (or emotions) for the origin of the judge's decision, but this fact is not sufficient for the decision itself. The judge is obliged to formulate his decision, and this fact is decisive for his *"legal"* life. If a judge relied on his or her legal intuition, or legal feeling, to justify a decision, it would be at least incomprehensible to the majority of the legal community, since a reasoning of judgement is a space for expressing explicit, logical reasons why the judge decided the case as stated in the verdict of the decision. In this respect, the reasoning of the judge's decision does not refer to the true cause of the decision (the judge's intuition) but is a rationalisation of his decision in the form of explicit legal reasons. Posner observes that the role of the subconscious in a judicial decision is "disguised" by the requirement of explicit justification, since we are supposed to justify an opinion that we have formulated intuitively by means of logical operations (Sobek, 2010, 177). Let us conclude the examination with Posner's words: *"The first decision in a line of cases may be the product of inarticulable emotion or hunch. But once it is given articulate form, that form will take on a life of its own – a valuable life that may include binding the author and the other judges of his court (along with lower-court judges) and thus imparting needed stability to law through the doctrine of precedent, though a death*

---

18 *"(...) the more experienced the judge, the more confidence he is apt to repose in his intuitive reactions and the less likely he is to be attracted to a systematic decision-making methodology"* (Posner, 2008, 108–109).

grip if judges ignore changed circumstances that make a decision no longer a sound guide" (Posner, 2008, 111).

The thoughts of all three authors we have discussed in this chapter have one thing in common: they emphasise the strong influence of intuition (respectively emotions) on the decision-making process of the judge, which is divided into two separate parts – the decision itself and the subsequent (*ex post*) rationalisation of the decision. But as we have already mentioned, Hutcheson, for example, did not believe that the intuitive decision-making in question was amenable to close scientific scrutiny; rather, it was the product of years of experience and experienced cases. But is it so? Today we know that it is not, and indeed several contemporary scholars in cognitive science and moral psychology have been studying the subject and coming up with interesting conclusions and practical implications for legal theorists and practitioners working in the field of law and public policy. So let's take a look at some of them.

Legal Intuition and Its Place in (Judicial) Decision-Making Process from the Perspective of Moral Psychology and Cognitive Science

While Cardozo, Hutcheson, Frank, and Posner are more or less relying on their own experience, or that of other judges and legal practitioners, other contemporary scholars working on this issue are formulating their conclusions based on the findings mainly from moral psychology, which is oriented towards the question of intuition in decision-making processes. Particularly since the 1990s, there has been a strong shift in moral psychology, which is also related to the desire of many not to place too much emphasis on rationalism and to deal more with issues of intuitionism. A large amount of literature produced within neuroscience, social psychology and primatology has shown us how powerful automatic and intuitive processes are (in particular, mention authors such as Haidt, Kahneman, Tversky and others). Haidt, in his book *The Righteous Mind*, describes this shift as "new synthesis" and presents the following three principles: (1) Intuitions come first, strategic reasoning second; (2) There's more to morality than harm and fairness; (3) Morality binds and blinds (2012, chapters 2-4). Given the scope of this chapter, the primary focus will be on the first principle, as it also has arguably the most direct implications for the legal community.

We approach the explanation of this principle in a similar way to Haidt (2006) or Kahneman (2011), using a metaphor: the mind is divided like a rider on an elephant, and the rider's job is to serve the elephant. The current view in psychology is that there are two basic and fundamentally different sorts of mental processes going on at all times in our minds: automatic processing (the elephant) and controlled processing (the rider). *"Most of human cognition is like that of other animals. All brains are neural networks, and they solve problems largely by pattern matching. This sort of process happens rapidly and automatically. When you open your eyes, you recognise objects and faces. You don't have to do any conscious work; your visual system just solves ferociously difficult computational problems nearly instantaneously, and it presents its results to your conscious awareness"* (Haidt, 2013, 869). This type of cognition has been familiar to mankind for tens of millions of years, but it is not the only way of knowledge. We also have a way of cognition that is unique to humans and consists of the use of language and words. Language, however, is probably no more than five or six thousand years old (Haidt, 2012, chapter 2). We have the ability to reason through language, but the process is slow and arduous. Similarly, if you think of situations when you are tired or drunk, the work becomes quite demanding and you make mistakes frequently.

Automatic cognition, symbolised by the great elephant, is sometimes referred to as "hot cognition". This is because it has the ability to nudge us into action; it is reactive and intuitive. It is not surprising to state that the neurological systems for judgement are connected to the neurological systems for behaviour (Haidt, 2013, 870). Controlled cognition, on the other hand, is referred to as "cool cognition" since it is not connected to the behavioural centres of the brain. And in this respect Haidt recommends us to imagine the human mind as a small and somewhat ineffectual rider perched on the back of a large, powerful, and rather smart elephant. *"The rider can try to steer the elephant, and if the elephant* has no particular desire to go one way or the other, it may listen to the rider. But if it has its own desires, it's going to do what it wants to do" (Haidt, 2013, 870).

Kahneman and his colleagues describe these processes a little differently and refer to the above ways of thinking as System 1 and System 2 (Kahneman, 2011, 19 et seq.; Kahneman and Klein, 2009, 515–526; for the legal reception of these ideas see also Berger, 2013; Richards, 2016). The first way of thinking (System 1, S1) is "fast" thinking and produces "effortlessly originating impressions and feelings that are the main sources of the explicit beliefs and deliberate choices of System 2" (Kahneman, 2011, 21). S1 is mainly characterised by low mental energy consumption and is used in most of our everyday activities and tasks, in solving which we use familiar and used thought processes, and these processes are subconscious and fast. *"S1 is a mark of our animal nature – its capabilities 'include innate skills that we share with other animals'. That doesn't mean we should discredit it, because it helps us go through various tasks and activities whose settlement are unproblematic and allows us to focus on the important and challenging aspects of our lives."* (Richards, 2016, 246). An adequate understanding of S1, however, is only possible in comparison to System 2 (S2), with S2 representing our "slower" thinking. In contrast to S1 *"the highly diverse operations of System 2 have one feature in common: they require attention and are disrupted when attention is drawn away"* (Kahneman, 2011, 22). It is necessary to understand that S2 not only costs us a lot of time and requires full concentration (which distracts us from our other activities), but it also costs us a lot of mental strength. *"You dispose of a limited budget of attention that you can allocate to activities, and if you try to go beyond your budget, you will fail"* (Richards, 2016, 246).

We should not neglect another characteristic of S2, namely that it is able to contain and compose two contradictive arguments in working memory. In contrast, S1 chooses a solution without pursuing other possible alternatives to the problem, or even admitting that there are other ways of solving it. Through S1, we are able to make quick decisions even without full information, as we are confident that we are doing the right thing while not giving too much weight to far-reaching consequences. Otherwise, if we consider every single decision we make, human action would be minimal. Therein lies the original characteristic of S2, namely laziness. S2 is lazy in the sense that this way of thinking does not occur if S1 seems to work fine (Kahneman, 2011, 31). It is a typical human trait to minimise energy expenditure as much as possible, even at the mental level. This means that S2 does not operate until certain stimuli arise, e.g., the person is confronted with a situation that is new to him or her or potentially life-threatening.

It is interesting to note, on the basis of the functioning of the human mind thus portrayed, that throughout the history of philosophy and thinking (and this includes the legal one) reason (i.e., the rider or S2) has been appreciated. This has been the case since Plato, who stresses the primacy of a priori knowledge (eternal ideas, the ideal world can only be known through rational activity). Plato himself, in the dialogue *Timaeus*, describes the myth about how the gods created man, beginning with the head, the seat of the immortal and rational soul, and then it was necessary to create the less perfect parts of the body, which were prone to succumb to the passions and emotions. Unsurprisingly, the drama of all life lies in the struggle of men's heads with their passionate bodies, with the intention of subordinating and directing the passions of the flesh to virtuous ends (Plato, 1998). Similarly, the modern rationalists, led by Descartes and Leibniz, worshipped reason (in a similar spirit to Plato) in the hope of modelling an entire philosophy and way of thinking on the deductive method developed by Euclid. Thus, even within moral philosophy, in moral reasoning, there are many claims that the rider should always have complete control over the elephant, i.e., the controlled cognition has control over the automated one (e.g., see Kohlberg, 1973). To put it another way: rationalist approaches in moral psychology say that moral knowledge and moral judgement are reached primarily by a process of reasoning and reflection (Kohlberg, 1969; Piaget, 1965). Intuition and moral emotions (such as sympathy, anger, guilt) may sometimes be inputs to the reasoning process, but they are not the direct causes of moral judgements. Reason and controlled cognition will eventually prevail.

However, Haidt (2013) argues, also based on a study of the writings of David Hume, that this is not the case. Reluctant to use similes as Hume did when he described reason as the slave of the passions (Hume, 2008) he rather interestingly chooses to compare reason as press secretary. "*The press secretary of a president serves the president, but it's a partnership. Her job is not to figure out the truth, or to make policy; it is to justify whatever the president and his cabinet have decided to do*" (Haidt, 2013, 871). The press secretary may have considerable influence on the running of the president and the "executive" and be a trusted advisor, but the president still runs everything. He or she will ultimately have to make the key decisions, but an important role of the press secretary is to develop and present arguments that portray the decisions in question to the public in the best light, in an effort to convince others of the justification for their implementation. Again, we come to the aforementioned claim: the decision and its justification are two separate processes, just as Hutcheson or Frank assumed in a judicial context.

On the basis similar to that presented by the American legal realists, Haidt subsequently comes up with The Social Intuitionist Model, which is based on the Humean model of intuitionism but supplemented with newer terminology and with an overlap into society. The central claim of the model is that moral judgement is caused by quick moral intuitions and is followed (when needed) by slow, *ex post facto* moral reasoning (Haidt, 2001, 817). The model itself is based on the following points:

1. the initial stage of the (moral) decision is guided by intuition (as hypothesised by Hutcheson or Frank in a judicial context), which is reactive to certain stimuli and eventually leads to a decision;
2. when a decision is reached, we then proceed to justify it, which provides an "apologia" for the judgement (we deliberately retain the term, which was used by Hutcheson); reasoning is an effortful process in which a person searches for arguments that will support an already-made judgement;
3. the decision and its justification may be subject to social interaction with another person, and this is often a significant part of *post-hoc* reasoning in order to get ready for the possibility that we might have to justify ourselves to others;
4. if we enter this debate with another person(s), we do so with pre-prepared reasons designed to convince the other side of the correctness of our view of the issue; we want them to "see it our way". However, if the opposing side has an intuitively differ-

ent position on the issue at hand, they typically do not change their minds; on the contrary, they prefer to come up with their own justification and reasons to confront us with, and the process then circles back around.
5. Of course, confronting our decision and its reasoning with the vision of the other party may also evolve in such a way that we change our perspective and reject our initial intuitive decision. Usually this is a situation where we are presented with new information of which we had no knowledge, and the introduction of this information by the opposing party may change our view of the matter under consideration.
6. Alternatively, there may sometimes be another situation, also mentioned by Frank, which may occur in deciding cases, where in *post-hoc* reasoning one may realise that the initial intuition was not correct and under the weight of producing a good justification for the decision, the initial wording of the decision is changed. Nevertheless, Haidt argues that these cases (of self-internal change of perspective) are probably fewer and do not occur that often in human thought; this may occur in cases in which the initial intuition is weak and processing capacity is high (Haidt, 2001, 819).

The social feature of this model is particularly interesting. Social intuitionist model proposes that moral judgement should be studied as an interpersonal process. According to Haidt, moral reasoning is usually an *ex post* facto process used to influence the intuitions (and hence judgements) of other people (2001, 814). When we reflect on some negative phenomenon (e.g., incest), we usually immediately (intuitively) feel that it is something wrong. However, when the societal need to justify that position is thrust upon us (especially when the case becomes more complex and complicated), even the legal layman subsequently has to put himself, as it were, in the role of a lawyer, to take up the "case" and build an argument around his feeling. It is questionable, however, with how much success, especially in interaction with others. And at the same time, if he runs out of arguments, subsequently the familiar comes up: "I don't know, I can't explain it, I just know it's wrong." In a difficult debate, the above-mentioned often foreshadows defeat in argumentation; however, within the social intuitionist model it becomes plausible to say it.

Awareness of the above processes can greatly assist in understanding what is really going on behind the decision-making processes of not only the judges but also of other persons in the position of law application bodies. Each of us is human and subject to certain determinants in our decision-making, and the intuitive ones can be crucial, while the ones proclaimed as logical, rational ways of reasoning in legal thinking may not be as dominant in our thinking as we have long thought. Before proceeding to the next part, let us try to summarise the main ideas arising from this chapter: we have the existing myth of rational, logically precise human decision-making. Many in the past have placed too much emphasis on rationality in cognition, but in fact we can see that intuition plays a significant role in human decision-making processes. Admittedly, some people continue to believe that there exists a reliable faculty of moral reasoning, capable of operating effectively and impartially even when self-interest, reputational concerns and intergroup conflict pull towards a particular conclusion. But Haidt (2013, 875) claims that no such faculty exists, and in understanding of the functioning of the human mind and its description, it is more appropriate to rely on authors like Hume and Hutcheson than on rationalists like Plato or Kohlberg.

Before proceeding to the concluding implications arising from the first two chapters of this text analysed above, we feel it is necessary to briefly discuss a concept that more closely describes the importance of intuitions, feelings and emotions in society and does not legitimately see them as something negative, unsystematic and subjective. This is the concept of moral emotions, which have a fundamental overlap with social life, as we will demonstrate with an interesting correlation: the feeling of injustice and the related moral (and social) emotion of anger.

The Feeling of Injustice and the Moral Emotion of Anger

It is interesting to note that many definitions of justice or injustice (as a grammatical antonym) by several legal theorists emphasise the importance of grasping this problem through the prism of intuition. Viktor Knapp, the eminent legal theorist, describes the concept of justice as *"intuitively comprehensible, but defining it causes considerable difficulty"* (Knapp, 1995, 86). Ota Weinberger, in turn, stressed that justice is not only a question of rationality but also a problem of emotionality and subjectivity, i.e., *"the pursuit of justice is a task of seeking, a task for the mind and the heart."* (Weinberger, 2010, 364). Even Hans Kelsen, perhaps the most important representative of twentieth-century continental legal positivism, thinks of justice as an irrational ideal, which on the one hand is indispensable for human will and behaviour but on the other hand is inaccessible to human cognition (Kelsen, 1933, 13). The irrational roots of justice for us were clearly expressed at the end of the twentieth century by Jiří Klabouch when he stated *"(...) the search for justice is not really about conflicts of ideas (...), but about the fluctuations of some very intense and deeply rooted emotions"* (Klabouch, 1995, 559). From our perspective, the contribution of Klabouch's theorem lies primarily in the distinction between the irrationality and the emotional force or significance (existential dimension) of justice attitudes and evaluations. These are important to human beings, and feelings of (in)justice are among the subjectively significant ones. At the same time, they do not emerge from the rationale, even though they are often later elaborated, articulated, organised and eventually enforced by it. Andrea Kluknavská (2021, 7) suggests that legal thinking and legal practice work with justice, but with it as an unspecified or undefined concept, despite the fact that jurisprudence usually carefully defines its notions (especially in the Slovak region, where normativist and dogmatic legal thinking still plays a significant role). This approach can be justified for a number of reasons, but two arguments in particular can be mentioned: (1) While legal thinking often uses the notion of justice, and it plays an important role in it, legal thinking respects that it is neither a legal nor a legal-philosophical notion. It leaves its definition primarily to social philosophy, from which legal theory draws when considering justice (Kluknavská, 2021, 7); (2) Justice is based on intuitions of what is (un)just, and in this sense escapes legal-theoretical knowledge (Bárány, 2011, 846).

In summary, for many legal thinkers of different backgrounds and orientations in our Central European territory, the irrational (extra-rational) basis of just ideas, evaluations and attitudes is an important (if not the key) component of justice. Bárány aptly notes that perhaps even the use of the concept of justice without definition, prevalent in legal reasoning, signals its extra-rational, intuitive basis for it is difficult to recognise in legal discourse that one of the fundamental concepts and arguments is rooted outside of rationality (2011, 851).

The above-mentioned also stems from the different understandings of justice. There is nothing new to say that justice takes on different meanings in different practical contexts, and to understand it fully one has to grapple with this diversity. Remaining in legal discourse, one of the most cited and accepted definitions of justice can be found in *Institutes of Justinian*, a codification of Roman Law from the sixth century AD, where justice is defined as *"the constant and perpetual will to render to each his due"*. It highlights important aspects of justice as such, and even on the basis of this short definition, one can understand the complexity and considerable vagueness of justice, since:

(1) Justice is significantly related to how particular people are treated ("to *each* his due"); typically, issues of justice arise in circumstances in which people can advance claims (to their rights, freedoms, opportunities, resources, etc.) but they can be potentially conflicting, and if so, we appeal to justice to resolve such conflicts by determining what each person is properly entitled to have.

(2) Justinian's definition underlines that just treatment is something *due* to each person; in other words that justice is a matter of claims that can be rightfully made against the agent dispensing justice, whether a person or an institution (Miller, 2021). However, there can be conflicts of virtues here: on the one hand, we have justice, which at times will also proclaim a severe response to man's wrongdoing (after all that is why the goddess Justice also has a sword in her hands); on the other hand, there is also a strong desire for forgiveness and love in our culture. To determine which virtue will have to prevail in a particular matter can be a rather tricky affair.

(3) An important aspect (related to legal thinking as well) is the relationship between justice and the impartial and consistent application of rules (the aspect in question refers to this part

of the definition: "constant and perpetual will"). Justice should be the opposite of arbitrariness. This aspect can be succinctly described in legal practice: if we have two identical or similar cases, their solution should be the same from the court's point of view. It emphasises two fundamental attributes of the application of the law, namely, the predictability of the law and its relative permanence (and at the same time, it is a rough expression of the principle of legal certainty). It is not surprising that this aspect is also intertwined with the concept of the rule of law: it is not to be governed by the arbitrary will of individuals or groups but by the rule of rational impersonal rules (Mrva, 2018, 99).

(4) Justinian's definition reminds us that justice requires an agent whose will alters the circumstances of its objects. An "agent" can be a wide range of entities: from an ordinary natural person to a smaller or larger group of people, to institutions (including the state). So we cannot, except metaphorically, describe as unjust states of affairs that no agent has contributed to bringing about (Miller, 2021). It must not be forgotten that this agency condition includes the possibility that injustice can also be committed by omission. In particular, this is so-called systemic injustice, for example, when policies and rules are not put in place to prevent structural discrimination against minorities (typically against racial, ethnic minorities or LGBTI+ communities).

As can be seen, legal thinking is significantly intertwined with the idea of justice, postulating the requirement of justice in law and its implementation (with an emphasis on the application of law), and at the same time reflecting the needs and functions of justice, where it distinguishes its several forms and divisions. Although it does not further define the notion of justice itself, despite this fact and the variety of approaches to justice, it is possible to extract a common understanding of it, certain shared foundations that are implicitly (and intuitively) present in it and legal thinking builds on them (Bárány, 2011, 846–847). We could, of course, continue and write at length in the above efforts to grasp justice.[19] How-

---

[19] One can also mention the way justice is understood through contrasting concepts, typically conservative vs. ideal justice, corrective vs. distributive justice, procedural vs. substantive justice, legal vs. ethical justice, comparative vs. non-

ever, we do not consider it necessary; with this short intermezzo we "only" wanted to point out three important moments, which are significantly related to the ideas already analysed in this chapter: (1) the diversity of understanding of the category of (in)justice in legal thinking; (2) its frequent seizing by intuitions and emotions, perceiving these as irrational (or extra-rational) elements; (3) at the end, it is the judge who has to decide on a particular case, and essential to that decision will be his or her perception of what is (un)just and which right or freedom he or she will grant or deny legal protection.

But let us move on and again relate the mentioned considerations to selected aspects of moral psychology. As we have pointed out, intuitions about what is (un)just towards whom, what goods and burdens are due to whom, are not just the application of explicit or implicit rules to a particular person and his situation. They are individual in their extra-rationality. But these intuitions should be investigated empirically, and the above is being done today, and it is a fascinating area of research. However, it should be noted that very rarely do legal theorists, philosophers or practitioners participate in these empirical studies. And while defining and describing justice can be found in almost every legal theory textbook (see, e.g., Knapp, 1995, 86-91; Prusák, 2001, 25-28; Gerloch, 2009, 251-253; Večeřa, et al. 2011, 310-317; Maršálek, 2018, 197-212), and many emphasise its intuitive, emotional or even irrational character (as we have shown above), this segment is understudied. This is surprising because:

a) justice and emotions are an important part of organisational life, characterising and informing organisational processes as well as acting as communication systems that help individuals navigate through the basic problems that arise in social relations;

b) and there is a strong theoretical association between (in)justice and emotion and from our perspective this is particularly evident in law and legal thinking.

At this point, we will not elaborate further on considerations

---

comparative justice, deontological vs. consequentialist justice, etc. See more on this in Schmidtz (2016); Miller (2011); Klabouch (1995); Weinberger (2017); and Fábry, Kasinec and Turčan (2019).

regarding the link between intuitions and moral emotions in general, and what the advantages of a given examination might be (this has been pointed out by several others, cf. Haidt, 2003, 2013; Horberg et al., 2011; Rudolph and Tscharaktschiew, 2014; Démuth, 2021; etc.). Exemplarily, however, we will take the liberty of pointing one relationship to show the usefulness of exploring this area, namely, the relationship between the feeling of injustice and the moral emotion of anger.

We know that experiencing justice results in positive emotions such as joy, pride, pleasure and so on (Weiss et al., 1999). However, injustice has been often described as a kind of hot and burning experience, and even third parties respond strongly to injustice (O'Reilly et al., 2016). The emotional experiences of (in)justice do not only affect attitude and behaviour at individual level but also affect intergroup attitude and behaviour at group level. Experiencing justice or injustice, however, is not usually understood as a moral emotion. Moral emotions are defined as those emotions *"that respond to moral violations or that motivate moral behaviour"* (Haidt, 2003, 853). So they are associated with social norms and values and the interests or welfare of persons or groups, beyond the concerns of the actors. That is, moral emotions function as signals of normative behaviour (see Hareli et al., 2013). Nowadays, there are many classifications of moral emotions. Eisenberg (2000) first proposed two kinds of moral emotions: self-conscious moral emotions (guilty, shame) and empathy. Lately, Haidt (2003) classified moral emotions into four kinds/families: self-conscious emotions (shame, embarrassment, and guilt), other-condemning emotions (contempt, anger, and disgust), other-suffering emotions (distress at another's distress, and sympathy) and other-praising emotions (gratitude, awe and elevation). We are currently encountering also this division of emotions: positive self-conscious moral emotion (proud), negative self-conscious moral emotion (shame, guilty), positive other-focused moral emotion (elevation, grateful), negative other-focused moral emotion (anger, disgust, contempt, compassion; see more in Horberg et al., 2011; Rudolph and Tscharaktschiew, 2014). Rudolph, Schulz, and Tscharaktschiew (2013) conducted a comprehensive literature search to identify those emotions that have been labelled as moral emotions by scientists. It showed that during the past 100 years, about two dozen emotion terms have been labelled as moral emotions. These are, in alphabetical order: admiration, anger, awe, contempt, disgust, elevation, embarrassment, empathy, envy, gratitude, guilt, indignation, jealousy, pity, pride, rage, regret, remorse, resentment, respect (including self-respect), *schadenfreude* (joy in the misfortune of others), scorn, shame and sympathy (compassion). But as we can see, neither justice nor injustice is mentioned here.

It is for this reason that perceived (in)justice is more often viewed as a triggering event (elicitor) for moral emotions. As we mentioned justice perception elicits positive moral emotions such as pride, pleasure and so on, and injustice perception activates strong negative emotions such as anger, disgust and contempt; besides, favourable unfair treatment triggers shame and guilt (Weiss et al., 1999; Rudolph and Tscharaktschiew, 2014; Tangney et al., 2007; Li et al., 2022). Significant in experiencing (especially) injustice is the aspect of actor/observer, or we can say first- and third-person perspective criterion. A wide range of studies suggest that people evaluate and respond not only to the (in)justice they personally experience but also to the (in)justice experienced by others (O'Reilly and Aquino, 2011; O'Reilly et al., 2016). When actors are involved in anger and frustration in injustice context, observers may have other considerations because injustice poses a threat to observers' social identity (Tyler and Boeckmann, 1997). They need to psychologically distance themselves from the deviant offenders through some behavioural responses like punishment or prosocial tendency (Lotz et al., 2011), suggesting that the offenders are not representative of their group and the values held by the majority. In addition, observers may fear that injustice will one day be done to them if justice is not restored in a timely manner, especially if observers and actors belong to an organisational system. Therefore, it is much possible for observers to experience higher levels of negative other-focused moral emotions in an injustice context. Further, the high level of negative other-focused moral emotions among observers and actors in conditions of injustice provides strong empirical support for the inequality aversion model and also explains well why the experience of injustice is "hot and burning" (Li et al., 2022).

The experience of injustice, described by the words "hot and burning", aptly foreshadows the most common moral emotion associated with it, namely anger, or even rage (Clayton, 1992; Smith et al., 1993; Haidt, 2003). A large number of philosophers in the past have described anger as irrational and tied to the subject who experiences it. Thus, it is a kind of asocialising moment in behaviour (Démuth, 2021).[20] However, several contemporary theorists point to the fact that anger as an emotion is primarily directed at someone else and is not as self-centred as it has been thought to be in the past. This is not only because we are often angered by others but especially in that our anger is directed at them. As we have already hinted above, there are several moral emotions that are oriented towards the subject (sadness, shame,...), i.e. towards the one who feels them. Such emotions are mostly not very outwardly manifested but are inwardly oriented. In contrast, however, there are also emotions that are primarily externally oriented. Their main addressee is the subject, but at the same time, the expression of the emotions experienced by the subject is oriented outwards – outside the subject – to someone else. Such emotions are, for example, joy, the expression of which is supposed to encourage the pleasantness of the stimulus and its desirability, but especially also anger or hostility, which are, on the contrary, warning signals of the individual in question and the expression of rejection of a certain stimulus, situation or phenomenon (Démuth, 2021). At the same time, one does not have to be a psychologist to see the anger signals in question, as Paul Ekman and Wallace V. Friesen (2003, 78–98) point out: signals such as furrowed eyebrows that pull together and down; narrowed to squinted eyes with a piercing and focused gaze, flanked from below by taut eyelids if the upper eyelids look drooped; clenched teeth reinforcing the entire face as if in a physical attack (or anticipation of a punch); or straining to control one's verbal expression or even mouth open in a scream with teeth bared, are clear signs of resentment and readiness to attack. Anger is meant to signal displeasure and a readiness to expend energy to remove that displeasure, thus the social aspect of the emotionality presented is evident.

---

20 The above aptly illustrates Seneca's claim about anger: "(...) *some wise men have said that anger is a brief madness: for it's no less lacking in self-control, forgetful of decency, unmindful of personal ties, unrelentingly intent on its goal, shut off from rational deliberation, stirred for no substantial reason, unsuited to discerning what's fair and true, just like a collapsing building that's reduced to rubble even as it crushes what it falls upon*" (Seneca, 2010, 14).

But let us not forget another aspect of this emotion that Aristotle (2007) had already pointed out when, in the second book of the Rhetoric, he describes anger as *"desire, accompanied by [mental and physical] distress, for apparent retaliation because of an apparent slight that was directed, without justification, against oneself or those near to one"* (Rhetoric, Book 2, Chapter 2). He thus moves away from the fact that one can be angry at oneself and at other things, but mainly thematises what we might call righteous/moral anger. The social and moral aspects of anger can be seen in the fact that one can often be angry at the wrongs and injustices done to others, and not just to oneself. But we also find it in the fact that Aristotle connects such anger with retribution and the idea of punishment as the levelling of wrongs (this is the well-known concept of corrective justice). Put differently, anger as a moral emotion has a mobilising character: it often emerges as a reaction to something in our environment that is disturbing to us. It is a moral emotion because it is preceded by an evaluation of the phenomenon, condition, situation or characteristic in terms of acceptability and desirability. However, its moral aspect also lies in the fact that it makes the other (the offender or the environment) clearly aware of the undesirability of a condition or behaviour and thus of the existence of possible negative consequences. When we are angry, we make our rejection clear. Since anger is one of the most "visible" and recognisable emotions, it is quite clearly readable to others. We make it clear to the disruptive elements that we are not satisfied with the *status quo*, that we want to change it, and that we are willing to invest a significant amount of our energy and time in doing so. Andrej Démuth (2021) also indicates that when we feel anger we are not only active, i.e. we make considerable efforts to change the behaviour of the person we are angry at (even, in escalated cases, with efforts to remove the "disruptor" itself), but in some cases the result of anger is to push the person out of our space, i.e., I become so offended at someone that we limit any contact with that person and consciously reject the potential benefits of possible mutual cooperation.

It is the experiencing of anger described above that makes it an important (also) evolutionary tool. Of course, we find cases where the tendency to act in anger may at first sight appear to be mainly selfish and antisocial, and indeed, in many cases it is. But it is not the only way of experiencing anger. We also have a strong motivation to correct injustice even in cases where it may be strongly felt in third-party situations in which one has no stake. Racism, mass human rights violations, exploitation of others and even ethnic cleansing or genocide can lead people who have no ties to the affected group to demand retaliatory or compensatory measures. In Radbruch's words, once we have experienced an intolerable degree of injustice, we are willing to invalidate a valid and effective law (Radbruch, 2013), as this fills us with righteous anger towards those who commit these wrongs. Similarly, today we see more frequently emerging and visible global expressions of anger towards certain negative phenomena. Due to social networking and easier communication across the world, we are more likely to encounter angry resistance to the diverse problems that plague our lives – from various social issues such as discrimination based on gender, race or sexual orientation (e.g., the well-known initiatives of Black Lives Matter or the #MeToo movement); through political challenges such as the increasing radicalisation and extremisation of society (the various manifestations of militant democracy); to the environmental difficulties of our planet (e.g., the well-known activities of Greta Thunberg against climate change). Public protests, riots, strikes and demonstrations of civil disobedience are just the consequences of the evolutionarily proven mechanism of expressing anger as a means to change the behaviour of those who have power but do not use it in accordance with our needs, desires or expectations. In this, we can see interesting and significant aspects of anger as a moral emotion for our society. Haidt (2003) even refers to anger as perhaps the most underrated emotion ever.

There is, however, one significant risk associated with anger, and that is that long-term ignoring of expressions of anger (even within the aforementioned activist manifestations) often leads to radicalisation of behaviour and the generalisation of anger to individuals or even entire nations. Indeed, an unpleasant peculiarity of anger is that it often, for the sake of simplicity, tends towards dispositional factors (Burton, 2018). If we are angered by an individual's behaviour, despite the fact that we are obviously sending out rejection signals demanding a change in behaviour, but it is not happening, we will be upset not only by the original behaviour but also by the individual's inability or unwillingness to modify their behaviour. The object of anger thus becomes not only the behaviour/guilt but also the refusal to change, or even the person itself. So often we move from situational anger to dispositional anger (i.e., hatred of the person, not just the situation caused by the person), which prompts us to the last possible solution: the removal of evil through the removal of the entire "evil" person from our world (Démuth, 2021). The above is dangerous in itself, all the more so if it "spreads" in society against a certain negative phenomenon, and perhaps here too we can see the driving force behind the various revolutions, which unfortunately have also brought about much bloodshed and death.

Of course, the understanding of the relationship between injustice and anger described above is a fragment of a possible examination. Truly, it is possible to perceive and see several interesting perspectives that can be related both to the feeling of (in)justice and our reactions to it, as well as to the emotion of anger, which in many ways not only helps us to move forward (namely at the level of the individual, the group and the society as a whole) but also brings with it many risks, because of which many well-known philosophers have preferred to accentuate the effort to master and control this hot and burning emotion.[21] We are not saying that this should accentuate and favour an angry way of dealing with problems in our private or public lives. By no means. Anger is not always desirable, quite the contrary. Nevertheless, at the same time, its fundamental impact on the life of each one of us and its mobilising function in society in the face of fundamental changes cannot be overlooked.

Simultaneously, it is necessary to understand that the emotion of anger (following the perception of injustice) will also ac-

---

21 Among all the concepts, let us mention, for example, the conception of *ataraxia*, in which it is essential to learn to control and manage individual emotions, and this is what often leads to the limitation of social tensions and conflicts. And this is what the Stoic social ideal leading to peace of mind is supposed to be. To reject evil, but without anger or emotion. To do it with apathy – without passion and with peace of mind. In this way, we should avoid conflict in society. However, it is questionable whether one is able to detach oneself from one's desires, emotions or feelings, and what is the right type of attitude each time, especially in the public sphere. We are not of this opinion, and at the same time, such an attitude suppresses several moral emotions and thus overlooks their importance and cognitive function in social life.

company legal thinking, including the sphere of the application of the law. We expect the courts to make just decisions, eliminating the injustices occurring in our society, in the absence of an important factor: anger. Judgements should never be passed in anger (or even rage), out of vengefulness or because of personal/collective hatred. Never. The construction of the decision itself, as well as its presentation in public, should meet the criteria of rationality and logic, sound legal reasoning and interpretation, as well as compliance with legal rules. But this is all part of the aforementioned *ex post* phase of rationalisation and justification of the judicial decision. However, our intuitive (automatic) feelings are often dominated by intuitions and emotions, including feelings of (in)justice and the often related emotion of anger. And from our point of view, this is significant precisely in eliminating the negative phenomena that plague us and society as a whole. It is impossible to ignore the fact that judges and others who function in the various positions of law application bodies are also human beings and in the event of a major social event, they too will be subject to the moods and emotions that flow through society. And often after the experience of injustice comes a righteous anger that cries out for some kind of change.

In Slovakia, we have had three such events in the last period in particular, which have "angered" the public. The murder of journalist Ján Kuciak and his fiancée Martina Kušnírová in 2018; the car crash at the Zochova bus stop in Bratislava in 2022, where a drunk driver killed five people; and the murder of two young people on Zámocká Street in Bratislava in 2022, conducted out of hatred towards sexual minorities. The majority of society felt both injustice and anger that the deaths had occurred. From our perspective, anger emerges for the very reason that certain types of behaviours or attitudes continue to occur that are not in line with society's desires and expectations. Political corruption and machinations in the exercise of public power; the arrogance to get behind the wheel in a state of drunkenness and endanger (and kill) others; or the hatred of others simply because they are of a different sexual orientation are all impulses that do not leave society "cold" and most of us have demanded, and continue to demand, change. And this change is also possible through law, and it has at least two levels:

- legislative: the legal system will be modified and adapted by the legislator to prevent, as far as possible, the recurrence of such negative events in society. In other words: the legal system also has to work with the anger that provokes in us resistance to selected types of unacceptable behaviour, since it is the legal system (specifically through the norms of criminal law) that signals that something is unacceptable, provokes resistance and has a tendency in the majority of society to actively reject it.
- the application: this is the reactionary level; if such wrongdoing has occurred, it must be justly investigated, verified and, at the end of the day, sanctioned. The sense of injustice and even anger that such sinister events have occurred at all is fully justified even among those who are in charge of applying the law. Undoubtedly, many judges, prosecutors, police officers, notaries, lawyers and other persons involved in the application of the law also feel these moods in society, and the above should not automatically be regarded as a bad phenomenon. Intuitive and emotional experience (even in the decision-making processes of individual subjects) helps to bring about change in society in this regard.

In conclusion, the perspective outlined above is in no way intended to "nudge" judges and others involved in the application of the law to make decisions and act only on the basis of gut feelings or emotional reasoning. The condition we have emphasised in the first two chapters remains relevant, namely that decision-making processes (including in law) must be seen in two steps: the making of the decision itself, where both the intuitive and emotional feelings of the decision-maker are relevant; and the subsequent (*ex post*) rationalisation of the decision in question. We believe that with the above-mentioned negative events that have happened in Slovakia in recent years, it is good if, when making decisions, judges and other actors who participate in the application of the law will also reflect the moods and emotions of society, they will influence them in their decision-making, and simultaneously, at the end of the day, judges and other actors themselves will contribute to a goal-oriented change in society and the elimination of those negative phenomena that burden the public. However, this is all in the presence of *ex post* rationalisation of their decisions and the need to defend their views and claims also within the legal discourse.

Conclusion

Surely, we are aware that our considerations touch on only a few moments regarding intuitive and emotional decision-making in legal thinking. With this chapter, we primarily wanted to contribute to a kind of demystification of law, with a particular emphasis on decision-making processes in law. In particular, we wanted to focus on the requirement of excessive rationality in the application of law that we often read in legal theory textbooks. Already the ancient designation of Roman law as *ratio scripta* (Rebro, 1995, 250), i.e. the understanding of law as written reason, is one of the many expressions of the self-evident fact that law should be rational, reasonable. Generally, to this day, the irrationality of law or its application is considered a deficiency in legal thinking. Interpretation of law and legal reasoning are systematically grounded on purposive rationality (Bárány, 2011). However, this chapter notes at least two levels where rationality can and does recede into the background in its application, and we do not consider this to be an automatic negative phenomenon. On the contrary, their awareness and subsequent reflection can help us in many ways to improve the functioning of law and its teaching at law schools.

We first addressed the issue of (legal) intuitionism in decision-making processes. Selected American legal realists emphasise to us the strong influence of intuition (respectively emotion) on the decision-making process of the judge, which is divided into two separate parts – the decision itself and the subsequent (*ex post*) rationalisation of the decision. In making their claims, they largely relied on their prior experience in legal practice, as well as some of them (Frank) worked with some already established information from other social science disciplines (especially psychology). Similarly, some experts in moral psychology share the above beliefs; in the presented text, we have given most attention to the thoughts of Jonathan Haidt and his conception of the social intuitionist model. However, both interdisciplinary approaches (the legal realist and Haidt's social intuitionist model) are inherently descriptive. They are descriptive claims, about how moral judgements are actually made. They are not normative or prescriptive claims, about how moral judgements ought to be made. Thus, if we accept the propositions presented in this chapter (which we believe it is possible to do), we believe that there are at least three basic implications for legal scholars and the legal community.

First: judges, like any other person in the decision-making process, are easily subject to extraneous and improper influences. There is an old lawyer's joke that the key to a good verdict by a judge is the proper timing of the hearing. To put it another way: how hungry a judge is greatly affects his or her decision-making processes, and for that reason, it is not a happy thing to have a hearing just before lunch. However, this joke has been partially confirmed several times in the past by various studies. Shai Danziger and his colleagues (2011) pointed out that it is relevant when judges ate their last meal when making decisions, and this is a determinant that can significantly influence judges' decision-making processes.[22] If we are aware of these processes, we are then better able to address the negative consequences (e.g., emphasising not overworking judges, as this can have a major impact on their decision-making processes).

Second: we consider that the theses of contemporary intuitionism (especially as presented by Jonathan Haidt) confirm selected ideas of Hutcheson and Frank. These legal realists correctly described decision-making processes by noting the diverse and extra-legal factors that influenced them; in particular, how intuition has a prominent place in these processes. Of course, this conclusion refers to the descriptive level of characterising application processes, not the normative one, which is again where the frequent criticism against legal realists is directed.[23] Jonathan Haidt (2003) opposes this view, arguing that empirical findings do not directly dictate normative conclusions, but one cannot engage in

---

[22] Their research shows that if you are first on the docket, right after the judges ate breakfast, you will probably be granted parole. But your odds decline drastically as the morning goes on, reaching pretty close to zero if you have the bad fortune to be the case heard just before the judges' late morning snack. After the snack your odds of parole shoot back up, to around 60%, but decline drastically again until the lunch break. Same story after lunch (Danziger et al., 2011).

[23] Many critics of legal realism still argue that if legal realism is to make a significant contribution to legal theory and legal philosophy, it should combine its critical, sceptical, and demythologising attitudes with normative and prescriptive ones, i.e., it should provide a guide to the making and interpretation of law in practice (Gábriš, 2020).

normative discussions until he/she has an accurate description of the kinds of creatures we happen to be. And legal realists were right about the kinds of creatures we happen to be.[24]

Third: the findings in question should gradually be transmitted to the academic life of law schools (this is particularly true in Slovakia). This conclusion is twofold: (1) legal science should no longer be exclusively carried out by professionals, i.e. lawyers; an interdisciplinary approach is necessary for a comprehensive understanding of the functioning of law and legal thinking, requiring the cooperation of several experts from diverse scientific fields;[25] (2) we believe that the education of lawyers should reflect the considerations we have described, as this can greatly assist in understanding the mechanisms of the law, as well as enable them to avoid selected negative phenomena and the consequences that are associated with it. Particularly relevant in this regard is clinical legal education, which has been called for in the past by some American legal realists (Frank, 1933; Meteňkanyč, 2020). From our point of view, it is essential that in addition to the study of positive law (also by using interactive methods), students should be given the opportunity to acquire knowledge from related social sciences, especially sociology and psychology. Students should be able to see, on the one hand, the interactions within the society where lawyers operate, as well as, on the other hand, to understand the "psychology" of the processes of law (especially the decision-making processes within the application of law) and that it is not a simple legal syllogism, but that several variables, including intuitive and emotional factors, play a role in these processes. Let's take one

---

[24] This was aptly summed up by another important representative of this movement, Oliver Wendell Holmes Jr. (1871), when he stated: *"The more we examine the mechanism of thought, the more we shall see that the automatic, unconscious action of the mind enters largely into all its processes. Our definite ideas are stepping-stones; how we get from one to the other, we do not know: something carries us; we do not take the step."*

[25] In this respect, this publication is also a certain attempt of interdisciplinary output, since the author publishes this paper as an outcome in a grant project supported by the Slovak Agency for Research and Development on the basis of the Contract no. APVV-19-0166, cooperating in this project with experts from the fields of psychology, philosophy, theology and law. Of course, it is also clear from the presented text that the author is a lawyer, but many of the impulses in the elaboration of this paper have come from colleagues, for which many thanks are due to them!

example for all which applies not only to law students but also to academics and legal practitioners, and that is how significant confirmation bias can be in our decision-making processes.

It is one of the most robust and deep-rooted biases in the literature of cognitive and social psychology. It's the finding that when we evaluate a proposition, we don't look for evidence on both sides and then weigh up which side is more likely to be true. Rather, we start with an initial hunch and then we set out to see if we can find any evidence to confirm it. If we find any evidence at all, we have confirmed the proposition, and we stop thinking. We believe that these types of heuristics and tendencies in (legal) thinking can be taught in a law school. And actually should be taught. Just think of the police investigators who have a hunch that the suspect is guilty. They will do everything in their power to confirm that hunch, and little to disprove it. Often they will arrive at false positive evidence that an innocent person is guilty. This is one of the reasons why it is so valuable to have an adversarial legal system – someone is appointed on each side to try to refute the other side's arguments. However, the above applies to all known legal professions as they often have decision-making competences. From judges, to notaries and bailiffs, to prosecutors. This is aptly described by Jennifer Lerner and Philip Tetlock (2003), who offer a unifying theory of judgement and decision-making that appears to be tailor-made for the legal community. They say our reasoning is heavily governed by accountability pressures. If you think that you might eventually be called on to explain yourself, you are going to reason much more carefully. But you are not going to work harder to figure out what's really true; you are going to reason much more carefully to figure out what is justifiable, what is defendable.[26] This seems to be applicable also in the case of judges, whose every written word may be scrutinised by an appellate court, but also by legal scholars, and by lawyers for interested parties.

The second issue that we have addressed more closely in this

---

26 "(...) a central function of thought is making sure that one acts in ways that can be persuasively justified or excused to observers. Indeed, the process of considering the justifiability of one's choices may be so prevalent that decision makers not only search for convincing reasons to make a choice when they must explain that choice to others, they search for reasons to convince themselves that they have made the 'right' choice" (Lerner and Tetlock, 2003, 433–434).

chapter, and which again points to the fact that too much emphasis should not be placed on rationality, is the influence of morally relevant feelings and emotions in legal thinking. We have exemplified the above-mentioned in the relation between the sense of (in)justice and the related moral emotion of anger. Legal thinking is predominantly built on rationality, but it also has to work with notions such as justice, which we have noted is intuitive and even irrational in nature, a fact that many legal theorists accept. Thus one encounters the interesting contradiction that justice connects law with irrationality, and the importance of the relation between law and the irrational (mediated by justice) lies in the existential power of some experiences and feelings of (in)justice. In the past, Bárány (2011) has pointed out that empirical research could identify which feelings of (in)justice are usually significant for a person and thus will be legally relevant. Frankly, this chapter is written in part as an initial effort to link these two perspectives, using the concept of moral intuitions and emotions (with an emphasis on the moral emotion of anger).

As we have mentioned, perceived (in)justice is viewed as an triggering event (elicitor) for moral emotions. Justice perception elicits positive moral emotions such as pride, pleasure and so on, and injustice perception activates strong negative emotions such as anger, disgust and contempt, besides, favourable unfair treatment triggers shame and guilt. The most common moral emotion that occurs when experiencing injustice is anger, which need not be seen as self-centred, subjective and negative. We also have a moral/righteous anger, tied to the wrongs and injustices done to us and to others, that is mobilising in nature. It often appears as a reaction to something in our environment that is disturbing to us, and the moral aspect also lies in making the other (the wrongdoer or the environment) clearly aware of the undesirability of a condition or behaviour, and thus of the existence of possible negative consequences. In this way, the moral emotion of anger can also be seen as an important evolutionary tool through which society (or a significant part of it) repudiates the negative phenomena occurring in it (today, most often global expressions of anger towards social or environmental issues). The emotion of anger thus portrayed (following the perception of injustice) will also accompany legal thinking, including in the sphere of the application of law.

In the text, we have tried to show how the experience of injustice and righteous anger can also be an impetus for change in law, whether at the legislative level, where the legal system also has to work with anger (and its social and moral aspects), as well as at the application level, where judges, prosecutors, police officers, notaries, lawyers and others who participate in the application of the law also feel these moods in society, and the above should not automatically be regarded as a bad phenomenon. We believe that intuitive and emotional experiencing (including in the decision-making processes of individual subjects) helps in this respect to achieve change in society; of course, in the presence of adhering to the condition that we have emphasised in the first two chapters, namely that one's initial intuitive or emotional decision in the application of legal relations must then be rationalised *ex post* and expressed in legal categories (i.e., on the basis of certain legal rules, analogies, concepts, and institutes). This should be done in such a way that it is legally defensible to a critical and receptive legal community and the wider public.

Obviously, we are not saying that rationalistic foundations in legal thinking are unnecessary or that they should be discarded. On the contrary, as we have suggested, law must continue to be composed based on them. What we have sought to demonstrate, however (and we have done so by combining two interdisciplinary concepts – selected aspects of legal realism and the recent concept of moral emotions), is how intuitive and emotional thinking can and does influence decision-making processes in law in important ways, and what interesting implications this may have. Too much adherence to the rationality of legal thinking distorts perceptions of the law, fails to capture its complexity and diversity, and even creates myths about the functioning of the law that do not reflect the reality of its application. And this should be avoided if possible.

References

Aristotle (2007). *On Rhetoric. A Theory of Civic Discourse* (Translated By George A. Kennedy). New York, Oxford: Oxford University Press.

Bárány, E. (2011). Spravodlivosť ako vzťahový pojem [Justice as a Relational Notion]. *Filozofia*, 66(9), 845–855.

Berger, L. (2013). A Revised View of the Judicial Hunch. *Legal Communication & Rhetoric: JALWD*, 10, UNLV William S. Boyd School of Law Legal Studies Research Paper Series, Available at SSRN: https://ssrn.com/abstract=2330532.

Burton, N. (2018). The Psychology and Philosophy of Anger. In: *Psychology Today* (Published On 09.12.2018). Available at: https://www.psychologytoday.com/us/blog/hide-and-seek/201812/the-psychology-and-philosophy-anger (accessed on 25.12.2022).

Cardozo, B., N. (2011). *Podstata súdneho procesu* [The Nature of the Judicial Process]. Bratislava: Kalligram.

Clayton, S., D. (1992): The Experience of Injustice: Some Characteristics and Correlates. In: *Social Justice Research*, 5(1), 71–91.

Cohen, F. (1935). Transcendental Nonsense and the Functional Approach. In: *Columbia Law Review*, 35(6), 809–849.

Colotka, P., Káčer, M., Berdisová, L. (2016). *Právna filozofia dvadsiateho storočia* [Legal Philosophy of the Twentieth Century]. Praha: Leges.

Danziger, S., Levav, J., Avnaim-Pesso L. (2011). Extraneous Factors in Judicial Decisions. In: *Proceedings of the National Academy of Sciences*, 108(17), 6889–6892.

Dea, S., Walsh, J., Lennon, T. M. (2018): Continental Rationalism. In: Zalta, E. N. (ed.). *The Stanford Encyclopedia of Philosophy* (Winter 2018 edition). available at: https://plato.stanford.edu/archives/win2018/entries/continental-rationalism/ (accessed on 25.12.2022).

Démuth, A. (2021): Hnev ako sociálna a morálna emócia [Anger as a Social and Moral Emotion]. In: Szakács, A. et al. (eds.). *Bratislavské Právnické Fórum 2021* [Bratislava Legal Forum 2021]. Bratislava: Univerzita Komenského v Bratislave, Právnická fakulta, 23–30. Available at: https://www.flaw.uniba.sk/fileadmin/praf/veda/konferencie_a_podujatia/bpf/2021/zbornik_bpf_2021_sekcia_1_teoria_prava_a_socialnych_vied_final.pdf (accessed on 25.12.2022).

Ekman, P., Friesen, W. V. (2003). *Unmasking the Face. A Guide to Recognizing Emotions from Facial Clues*. Los Altos, CA: Malor Books.

Fábry, B., Kasinec, R., Turčan, M. (2019). *Teória práva* [Theory of Law]. Bratislava: Wolters Kluwer.

Frank, J. (1933). Why Not a Clinical Lawyer-School? In: *University of Pennsylvania Law Review*, 81(8), 907–923.

Frank, J. (1948). Say It with Music. In: *Harvard Law Review*, 61(6), 921–957.

Frank, J. (1949). *Law and Modern Mind*. London: Stevens & Sons Limited.

Gábriš, T. (2020). *Preskriptívna teória práva: metodológia aplikácie práva pre súčasnosť* [Prescriptive Theory of Law: A Methodology of the Application of Law for the Present]. Bratislava: Veda.

Gerloch, A. (2009). *Teorie práva* [Theory of Law]. Plzeň: Aleš Čeněk.

Haidt, J. (2001): The Emotional Dog and Its Rational Tail: A Social Intuitionist Approach to Moral Judgment. *Psychological Review*, 108(4), 814–834.

Haidt, J. (2006): *The Happiness Hypothesis: Finding Modern Truth in Ancient Wisdom.* London: Arrow Books.

Haidt, J. (2012): *The Righteous Mind: Why Good People Are Divided by Politics and Religion.* New York: Pantheon Books.

Haidt, J. (2013). Moral Psychology and the Law: How Intuitions Drive Reasoning, Judgment, and The Search For Evidence. *Alabama Law Review,* 64(4), 867–880.

Hareli, S., Moran-Amir, O., David, S., Hess, U. (2013). Emotions as Signals of Normative Conduct. *Cognition & Emotion,* 27(8), 1395–1404.

Harvánek, J. et al. (2008). *Teorie práva* [Theory of Law]. Plzeň: Aleš Čeněk.

Holmes, O. W. (1871). *Mechanism in Thought and Morals.* Available at: https://hackneybooks.co.uk/books/454/777/mechanismthoughmorals.html (accessed on 25.12.2022).

Horberg, E., J., Oveis, C., Keltner, D. (2011). Emotions as Moral Amplifiers: An Appraisal Tendency Approach to the Influences of Distinct Emotions upon Moral Judgment. *Emotion Review,* 3(3), 237–244.

Hume, D. (2008). *A Treatise of Human Nature.* Wilmington: Nuvision Publications (Original Work Published 1739).

Hutcheson, J. C. (1929). Judgement Intuitive: The Function of the "Hunch" in Judicial Decision. In: *Cornell Law Quarterly,* 14(3), 274–288.

Kahneman, D., Klein, G. (2009). Conditions for Intuitive Expertise: A Failure to Disagree. *American Psychologist,* 64(6), 515–526.

Kahneman, D. (2011). *Thinking, Fast and Slow.* New York: Farrar, Straus And Giroux.

Kelsen, H. (1933). *Ryzí nauka právní. Metoda a základní pojmy.* [Pure Theory of Law. Method and Basic Concepts]. Brno – Praha: Orbis.

Kelsen, H. (2018). *Čistá právna náuka* [Pure Theory of Law]. Bratislava: Kalligram.

Klabouch, J. (1995). Pluralistické výklady spravedlnosti [Pluralistic Interpretations of Justice]. *Právník,* 134(6), 558–562.

Kluknavská, A. (2021). *Spravodlivosť v dejinách právneho, politického a sociálneho myslenia (od počiatku dejín myslenia po myšlienky K. Marxa)* [Justice in the History of Legal, Political and Social Thought (From the Beginning of the History of Thought to the Ideas of K. Marx)]. Bratislava: Univerzita Komenského, Právnická Fakulta.

Knapp, V. (1995). *Teorie práva* [Theory of Law]. Praha: C. H. Beck.

Kohlberg. L. (1969). Stage and Sequence: The Cognitive-Developmental Approach to Socialization. In: Goslin, D. A. (ed.). *Handbook of Socialisation Theory and Research.* Chicago: Rand Mcnally, 347–480.

Kohlberg, L. (1973). The Claim to Moral Adequacy of a Highest Stage of Moral Judgment. *The Journal of Philosophy,* 70(18), 630–646.

Langer, S. (1948). Philosophy in a New Key: A Study in the Symbolism of Reason, Rite, and Art. *The New American Library.* Available at: https://monoskop.org/images/6/6c/langer_susanne_k_philosophy_in_a_new_key.pdf (accessed on 25.12.2022).

Lerner, J., Tetlock, P. (2003). Bridging Individual, Interpersonal, and Institutional Approaches to Judgment and Decision Making: The Impact of Accountability on Cognitive Bias. In: Schneider, S., Shanteau, J. (eds.). *Emerging Perspectives on Judgment and Decision Research*. Cambridge: Cambridge University Press, 431–457.

Li, X., Hou, M., He, Y., Ma, M. (2022). People Roar at the Sight of Injustice: Evidences from Moral Emotions. *Current Psychology*, Available at: https://doi.org/10.1007/s12144-022-04014-w (accessed on 25.12.2022).

Lotz, S., Okimoto, T., G., Schlösser, T., Fetchenhauer, D. (2011). Punitive Versus Compensatory Reactions to Injustice: Emotional Antecedents to Third-Party Interventions. *Journal of Experimental Social Psychology*, 47(2), 477–480.

Markie, P., Folescu, M. (2021). Rationalism vs. Empiricism. In: Zalta, E. N. (ed.). *The Stanford Encyclopedia of Philosophy* (Fall 2021 Edition). available at: https://plato.stanford.edu/archives/fall2021/entries/rationalism-empiricism/ (accessed on 25.12.2022).

Maršálek, P. (2018). *Příběh moderního práva* [The Story of Modern Law]. Praha: Auditorium.

Meteňkanyč, O. M. (2018). Právny intuicionizmus [Legal Intuicionismus]. In: Kluknavská, A., Gábriš, T. (eds.). *Mýty o práve, mýty v práve* [Myths about Law, Myths in Law]. Bratislava: Wolters Kluwer, 206–222.

Meteňkanyč, O. M. (2020). Právnické vzdelávanie v podaní J. Franka a aktuálnosť jeho odkazu pre dnešné právnické fakulty [Legal Education in the Work of J. Frank and The Relevance of his Legacy for Today's Law Schools]. In: Szakács, A., Hlinka, T. (eds.). *Bratislavské právnické fórum 2020* [Bratislava Legal Forum 2020]. Bratislava: Univerzita Komenského v Bratislave, Právnická Fakulta, 71–81. available at: https://www.flaw.uniba.sk/fileadmin/praf/bpf/2020/zborni__k_teorka_2020_-_final_ocislovane.pdf (accessed on 25.12.2022).

Miller, D. (2021). Justice. In: Edward N. Zalta (ed.). *The Stanford Encyclopedia of Philosophy* (Fall 2021 Edition). available at: https://plato.stanford.edu/archives/fall2021/entries/justice/ (accessed on 25.12.2022).

Modak-Truran, M. C. (2001): A Pragmatic Justification of the Judicial Hunch. *University of Richmond Law Review*, 35(1), 55–89.

Mrva, M. (2018). Demokracia a právny štát [Democracy and the Rule of Law]. In: Kolektív autorov [Collective of authors]. *Aktuálne otázky teórie práva* [Current Issues in the Theory of Law]. Bratislava: Wolters Kluwer, Univerzita Komenského, Právnická fakulta, 84–102.

O'Reilly, J., Aquino, K. (2011). A Model of Third Parties' Morally Motivated Responses to Mistreatment in Organizations. *The Academy of Management Review*, 36(3), 526–543.

O'Reilly, J., Aquino, K., Skarlicki, D. (2016). The Lives of Others: Third Parties' Responses to Others' Injustice. *Journal of Applied Psychology*, 101(2), 171–189.

Paul, J. (1959). *The Legal Realism of Jerome N. Frank. A Study of Fact-Skepticism and the Judicial Process*. Hague: Martinus Nijhoff.

Piaget, J. (1965). *The Moral Judgement of the Child* (Translated by M. Gabain). New York: Free Press (original work Published 1932).

Plato (1998). *Timaeus* (Translated by Benjamin Jowett). Available at: https://www.gutenberg.org/files/1572/1572-h/1572-h.htm#link2h_sect1 (accessed on 25.12.2022).

Posner, R. (2008). *How Judges Think*. Cambridge: Harvard University Press.

Powell, H. J. (2009). A Response to Professor Knight. *Duke Law Journal*, 58, 1725–1729.

Prusák, J. (2001). *Teória práva* [Theory of Law]. Bratislava: Vydavateľské oddelenie Právnickej fakulty Univerzity Komenského.

Prümm, H. P. (2016). The Didactic Turn of German Legal Methodology. *Jurisprudencija: Mokslo Darbu Žurnalas*, 23(2), 1233–1282.

Radbruch, G. (2013). *O napětí mezi účely práva* [On the Tension between the Purposes of Law]. Praha: Wolters Kluwer.

Rebro, K. (1995). *Latinské právnické výrazy a výroky* [Latin Legal Terms and Sayings]. Bratislava: Iura Edition.

Richards, D. (2016). When Judges Have a Hunch. Intuition (and Some Emotion) In Judicial Decision Making. *Archiv für Rechts-und Sozialphilosphie*, 102(2), 245–260.

Rudolph, U., Schulz, K., Tscharaktschiew, N. (2013). Moral Emotions: An Analysis Guided by Heider's Naive Action Analysis. *International Journal of Advances In Psychology*, 2(2), 69–92.

Rudolph, U., Tscharaktschiew, N. (2014). An Attributional Analysis of Moral Emotions: Naïve Scientists and Everyday Judges. *Emotion Review*, 6(4), 344–352.

Schmidtz, D. (2016). *Prvky spravodlivosti* [The Elements of Justice]. Praha: Dokořán.

Seneca, L. A. (2010). *Anger, Mercy, Revenge* (Translated by Robert A. Kaster and Martha C. Nussbaum). Chicago, London: The University Of Chicago Press.

Smith, C. A., Haynes, K. N., Lazarus, R. S., Pope, L. K. (1993). In Search of the "Hot" Cognitions: Attributions, Appraisals, and Their Relation to Emotion. *Journal of Personality and Social Psychology*, 65(5), 916–929.

Sobek, T. (2010). *Nemorální právo* [Immoral Law]. Praha: Ústav Státu A Práva.

Sobek, T. (2011). *Právní myšlení. Kritika moralismu* [Legal Thinking. Critique of Moralism]. Praha: Ústav státu a práva.

Stępień, M. (2019). Say It with Images: Drawing on Jerome Frank's Ideas on Judicial Decision Making. In: *International Journal for the Semiotics of Law – Revue Internationale De Sémiotique Juridique*, 32, 321–334.

Tangney, P. J., Stuewig, J., Mashek, D. J. (2007). Moral Emotions and Moral Behavior. *Annual Review of Psychology*, 58(1), 345–372.

Turčan, M. (2020). K téme "dogiem" v práve [On the Subject of "Dogmas" in Law]. In: Szakács, A., Durec Kahounová, M., Senková, S. (eds.). *Míľniky práva v stredoeurópskom priestore 2020* [The Milestones of Law in the Area of Central Europe 2020]. Bratislava: Univerzita Komenského v Bratislave, Právnická fakulta, 13-19. available at: https://www.flaw.uniba.sk/fileadmin/praf/milniky/2020/milniky_zbornik_2020_final.pdf (accessed on 25.12.2022).

Tyler, T. R., Boeckmann, R. (1997). Three Strikes and You Are Out, but Why? The Psychology of Public Support for Punishing Rule Breakers. *Law and Society Review*, 31(2), 237-265.

Večeřa, M. et al. (2011). *Teória práva* [Theory of Law]. Bratislava, Žilina: Paneurópska vysoká škola, Eurokódex.

Volkomer, E. W. (1970). *The Passionate Liberal: The Political and Legal Ideas of Jerome Frank*. Hague: Martinus Nijhoff.

Weinberger, O. (2010). *Inštitucionalizmus: Nová téoria konania, práva a demokracie* [Institutionalism: A New Theory of Action, Law and Democracy]. Bratislava: Kalligram.

Weinberger, O. (2017). *Norma a instituce – úvod do teorie práva* [Norms and Institutions – Introduction to the Theory of Law]. Plzeň: Aleš Čeněk.

Weiss, H., M., Suckow, K., Cropanzano, R. (1999). Effects of Justice Conditions on Discrete Emotions. *Journal of Applied Psychology*, 84(5), 786-794.

Weyr, F. (2015). *Teorie práva* [Theory of Law]. Praha: Wolters Kluwer.

Williams, B. (1967). Rationalism. In: Edwards, P. (ed.). *The Encyclopedia of Philosophy* (Vol. 7-8). New York: Macmillan, 69-75.

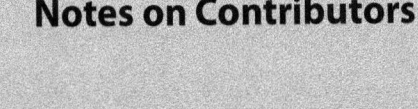

Ľubomír Batka is Professor at the Department of Theory of Law and Philosophy of Law, Faculty of Law, Comenius University in Bratislava. He earned his PhD at Eberhard-Karls Universität in Tübingen, in field of Protestant Theology and habilitated at Comenius University in Bratislava. He was visiting Fellow at Leibniz Institute for European History in Mainz. Currently he works and publishes in ethics and bioethics.

Andrej Démuth studied philosophy and psychology. He is Professor of Philosophy and Head of the Center for Cognitive Studies at the Department of Theory of Law and Philosophy of Law, Faculty of Law, Comenius University in Bratislava. He is the author of many books and articles on cognition and the relationship between reflected and non-reflected knowledge, and he regularly gives invited lectures at universities in Slovakia and abroad. His research focuses on modern philosophy, epistemology and cognitive studies.

Slávka Démuthová is Professor of Psychology and a Head of the Centre for the Psychological Counselling and Research at the Faculty of Arts, University of Ss. Cyril and Methodius in Trnava, Slovakia. Her professional orientation focuses on the developmental problems of children and youth as well as on the biological/evolutionary explanations of human behaviour. She is an author of several monographs, and scientific articles and regularly gives invited lectures at universities abroad (Edinburgh, Brno, Prague, Warsaw, Ljubljana).

Yasin Keceli studied psychology (BA – Aydin) and cognitive studies (MA – Trnava). He is a PhD. student in Systematic Philosophy at the Department of Philosophy and Applied Philosophy at the Faculty of Arts, University of Ss. Cyril and Methodius in Trnava. His PhD. thesis is focused on Cross-Cultural Comparative Research on Aesthetical Experiences of Beauty.

Renáta Kišoňová studied philosophy at Trnava University. She works as Lecturer at the Department of Theory of Law and Philosophy of Law, Faculty of Law, Comenius University in Bratislava. Her research interests include the problems of cognitive aesthetics, mainly the issue of facial and portrait's perception She has also published a monograph "Faces of a face. Portraiture as a means of representing faces". Currently, she focuses on the problem of social and aesthetical emotions of disgust and admiration.

Olexij M. Meteňkanyč studied law and philosophy. He is Assistant Professor at the Department of Theory of Law and Philosophy of Law, Faculty of Law, Comenius University Bratislava. He is the author and co-author of numerous scientific articles and textbooks, and in his academic activities he is not only addressing general theoretical, legal-philosophical and ethical issues but also focuses specifically on the topic of justice, both at the legal-philosophical level (in particular in the recent debates between representatives of iusnaturalism and iuspositivism) and in human rights sphere (especially the protection of the human rights of minorities). He regularly participates in national and international conferences, seminars and workshops.

www.ingramcontent.com/pod-product-compliance
Lightning Source LLC
Chambersburg PA
CBHW052100300426
44117CB00013B/2225